D0214632

STRAND BOOK STORE

APOCALYPSE AND
PARADIGM

APOCALYPSE AND PARADIGM
Science and Everyday Thinking

Errol E. Harris

Westport, Connecticut
London

Library of Congress Cataloging-in-Publication Data

Harris, Errol E.
    Apocalypse and paradigm : science and everyday thinking / Errol E.
Harris.
        p.   cm.
    Includes bibliographical references and index.
    ISBN 0–275–96830–8 (alk. paper).
    1. Philosophy and science.   2. Science—Social aspects.
3. Science—Philosophy.   I. Title.
    B67.H372   2000
    501—dc21   99–43108

British Library Cataloguing in Publication Data is available.

Copyright © 2000 by Errol E. Harris

All rights reserved. No portion of this book may be
reproduced, by any process or technique, without the
express written consent of the publisher.

Library of Congress Catalog Card Number: 99–43108
ISBN: 0–275–96830–8

First published in 2000

Praeger Publishers, 88 Post Road West, Westport, CT 06881
An imprint of Greenwood Publishing Group, Inc.
www.praeger.com

Printed in the United States of America

The paper used in this book complies with the
Permanent Paper Standard issued by the National
Information Standards Organization (Z39.48–1984).

10   9   8   7   6   5   4   3   2   1

Contents

Preface

At the beginning of the third millennium of the Christian era, the human race is menaced by terminal ailments that demand immediate and comprehensive treatment, of which there is as yet little sign and scant prospect. The nature and the seriousness of the dangers has long been recognized, and scientists have been warning the public and politicians for decades of the deterioration of the global environment and its fearsome results. It is widely accepted that this environmental deterioration has been and is still being caused largely by human activity: by the emission of industrial wastes and of greenhouse gases that produce global warming; the pollution of rivers, lakes, and oceans; the destruction of tropical rain forests; and over-use of fertilizers and insecticides and the general overcultivation of land which causes soil exhaustion and erosion, exterminates useful and scientifically important species, destroys habitats, and disrupts ecological systems. Still, nothing approaching adequate steps have been taken to check the processes of destruction nor to counteract their effects. One is immediately prompted to ask why.

It is a well-established fact that global warming is proceeding apace and that human activities are contributing significantly to its causes, but such measures as are contemplated to limit the use of fossil fuels and to restrain emissions of greenhouse gases are gravely insufficient and lamentably unreliable.

The climate change that will result is no trivial matter. Impressive evidence is already available of the break-up of the polar icecaps. The number of icebergs breaking off and floating south from the Arctic ice sheet have significantly increased in recent years, and a similar process has begun on the margins of Antarctica. As they drift into ever lower latitudes they are likely to alter the ocean currents, diverting the gulf stream to render the climate of northwestern Europe similar to that of Labrador and northern Canada, while currents like El Niño cause drought and dessication in Southeast Asia.

The melting icecaps will raise the sea level to submerge large portions of the coasts of the continents and many low-lying islands, all of which at present are densely populated. Global warming also portends more severe weather patterns, more violent and more widespread hurricanes, larger and more powerful tornadoes, bringing massive death and destruction to tropical and even to presently temperate regions.

In these circumstances one would expect a serious and determined effort to adopt measures to prevent the emission of greenhouse gases, such as the redesign of the internal combustion engine, enabling it to use cleaner fuels, the more widespread use of renewable energy, the prevention and more efficient disposal of industrial waste, the prohibition of felling and burning of tropical rain forests and action to increase their growth and spread to repair damage already done. But none of these things are being seriously contemplated. In fact, destructive practices that have already proceeded much too far are continuing unchecked.

Global warming is only one of the major threats to human survival prevailing at the present time; the effects of ecological disruption and the calamitous loss of species currently occurring threaten to break the food chain and forebode starvation to millions of people, posing serious threats to human survival. And besides all this, the persistent failure of international organizations to eliminate war, ethnic cleansing (as in Bosnia and Kosovo), internecine massacres (as in Rwanda), and the continued menace of nuclear holocaust (whether caused by accident, terrorism, or deliberate military strategy) is not being countered by any serious attempt to curb the excesses of nationalism, or to address the problem of national sovereignty, the prevailing obstacle to the maintenance of peace and a worldwide rule of law.

My main concern, however, is not to alarm the reader with gruesome descriptions of these menaces, but to investigate the reasons for the general failure to encounter them.

Human beings are intelligent animals. When humans are warned of dangers such as those that now overshadow them, one would expect them to take appropriate action. Yet they have not and do not; nor do they show any clear signs of awareness of the impending calamity, although the relevant information is easily and quite copiously available. What has come over people to cause them to behave like the Gadarene swine rushing blindly into the abyss of disaster?

It must puzzle the thoughtful observer that, although scientists know quite well what the dangers are of the present situation, as well as what can and ought to be done to meet them, and although they have made this information generally available, there is so little public reaction to their disclosure. Scientists cannot themselves take the necessary action, for this requires legislation that they have no power to enact or enforce. It requires political action and what nowadays is referred to as political will. That is what seems to be lacking, although the politicians and the public have been apprised of the facts and are aware of the available remedies. What is the cause of this paralysis?

The facts that the perils looming over the civilized world are known to science and have been imparted to politicans and that yet these politicians show only a slight inclination to take sufficient action, seem to indicate that there is somehow a discrepancy between the thinking of the scientists and that of the lay public. It is my object to seek the source and nature of this discrepancy and to consider how, if at all, it might be overcome.

It has occurred to me that the reason must lie in inveterate habits of thinking and acting which preclude the appropriate response to the obvious dangers. The kind of action needed is clearly dependent on scientific knowledge: scientific assessment of the deleterious processes currently taking place, and scientific methods of preventing or reversing them. The prevailing conceptual scheme dominating and controlling scientific research must be of direct relevance. I have therefore decided to examine the nature of scientific paradigms and their influence on the thinking and practice not simply of scientists but of those operating in other realms of civilized life.

Science is an outgrowth from civilized culture, dependent on other prior intellectual developments and concurrent with social and political activities of which it is not independent and which are not themselves uninfluenced by scientific thinking. The interrelation and interaction of these cultural activities and possible discrepancies between them may well account for the ineffectual conduct of peoples and their leaders in the present crisis, and the effect of scientific concepts on other activities will not be insignificant.

What follows is an attempt to consider how, in the past and in the current century, scientific paradigms have influenced thinking in philosophy and morality, economics and politics; as well as the extent to which this has affected people's attitudes and actions in the past and may be affecting our conduct in the predicament that faces us today. My submission is that the conceptual scheme introduced by the Copernican revolution in the sixteenth century permeated and transmuted philosophical, ethical and political thinking and correlatively social practice; and that although a new scientific revolution has taken place in our own day, it has so far failed to penetrate thinking and practice in other fields of intellectual activity. This, I believe, is the source of our present predicament.

The troubles that beset the human race at the turn of the century are all consequent upon the technical inventions spawned by science; and the use that has been made of them has been prompted by attitudes to the world and the environment engendered by the science born of the Copernican revolution. Now, however, that revolution has been superseded by the Einsteinian revolution in physics, which has drastically altered the world picture. Yet there is plentiful evidence that current habits of thinking in most if not all spheres of activity other than science are still those typical of the earlier scientific paradigm.

Because of this delay on the part of moral, political, and economic thinking to catch up with that of physics and biology, it is probable that the scientists are able to see the current state of affairs more clearly than the politicians and administrators can discern the appropriate action to take to set matters right. What remains to be decided is the way out of the consequent dilemma in which we find ourselves, and this may prove to be even more difficult to discover than the reason for the lack of adequate response to the obvious dangers.

1

The Significance of
Conceptual Schemes

What has come to be called a scientific paradigm is more far
reaching and influential than is usually recognized. Scientists
seek to understand and explain the phenomena they observe,
and they do so against the presumed background of an orderly
world, the general nature of which they tacitly presuppose. This
presupposition is part and parcel of their "paradigm," but it is
not exclusively scientific (in the current and narrower sense of
that word), for it involves a metaphysic which invades and is
latent in other spheres of reflective thinking. The implications
of a prevailing paradigm for, and its effects upon, other human
activities than science are therefore very wide ranging and can
have consequences of vital importance to the contemporary
civilization.

The word "paradigm" is defined by the *Oxford English Dic-
tionary* as: "example, pattern, especially of inflexion of noun,
verb, etc.," but the term was appropriated by Thomas Kuhn to
mean "research firmly based on one or more past scientific
achievements . . . that some scientific community acknowledges

for a time as supplying the foundation for its further practice."[1]
It transpires, as he goes on to elaborate what he means by the
term and explain the way he uses it, that it connotes and includes the conception of the basic elements of physical reality
and the overall view of the universe tacitly assumed by the
scientific community concerned. What Kuhn calls a paradigm
is, in effect, what Collingwood called the constellation of absolute presuppositions of science in a particular historical period,
absolute presuppositions being what give rise to scientific questions and are not themselves answers to such questions.[2] This
constellation of absolute presuppositions, he said, was the task
of metaphysics to discover, by logical analysis of scientific assertions. In fact, the absolute presuppositions of science turn
out to be the metaphysical doctrine that constitutes its unac-
knowledged foundation. What these thinkers have in mind is
the presumed world view characteristic of a particular historical period or epoch, what I prefer to call the conceptual scheme
fundamental to systematic thinking at that time.

The use of the term "paradigm," inspired by Kuhn, has now
become customary and is so common that, as long as it is understood in the way I propose, it will be convenient to adopt it
when referring to such a conceptual scheme. But we must recognize that, although the paradigm is established mainly by
science and made explicit by philosophers, it is basic to a form
of civilization and affects every facet and aspect of civilized
life in the given epoch.

Few will deny that the outstanding characterstic of contemporary civilization is the dominating position of science. In one
form or another, science permeates everything people do all
over the world. Science cares for our health, and its technical
inventions examine our bodies and test our vital functions. Our
pastoral and agricultural methods are governed by technology
that science has generated. Scientific appliances wash our
clothes and our dishes, sweep our floors, heat our homes, and
cook our food. Scientific inventions provide our entertainment,
our means of communication, and our transport. They impart
our news and information. They write our books and convey
our correspondence. At one extreme they invade our most inti-

mate activities, and at the other they explore outer space. We revere scientists as the most reliable of savants. We look to them for the solutions to all our problems.

So prevalent and authoritative has science become that we tend to think of it as a self-contained, self-sufficient activity, offering explanation of puzzling phenomena and the "know-how" for complicated technology and machinery that serves virtually all our needs. In fact we tend to equate science with civilization itself, considering peoples who lack it, or are scientifically unsophisticated as, to that extent, uncivilized, and those who enjoy its advantages as the most advanced. But this is a serious misapprehension: The very existence of science depends on the prior development of a civilized culture and not the contrary; science is closely interwoven with other factors contributing to orderly and regulated living.

Western European civilization is more markedly scientific than that of the East or than any other civilizations, wherever they may have arisen. Certainly, many scientific inventions occurred in China before they were introduced into Western Europe, but Chinese science has never been as systematic nor, independently and apart from Western influence, has it advanced as spectacularly as it has in Europe and the United States. But Western science itself is not self-contained. It has a long history and emerged from other concomitant activities contributing to civilized life.

In the first place no science is possible apart from an organized social structure and a culture including religion and art. It cannot develop without a disciplined educational practice, which itself presupposes the existence of an organized social system. In the West it grew out of early imaginary mythological legends designed to explain natural occurrences. These prompted and engendered scientific speculation.[3] Religion and science are therefore not separate intellectual pursuits. Both claim truth, and thus are mutually related, whether in harmony or in conflict. Religion originates in the awe inspired by natural occurrences to which human life is subject, and religious mythology derives from human wonder and the attempt to explain the major cycles and events in nature. Concurrently, philosophy and science have always been closely connected and were, in fact, originally one and the same. It follows that, just as science depends on other civilized and civilizing processes, so the other features of culture are affected by scientific theories and discoveries. None of these cultural factors is completely isolable from any of the others.

Such science as there was in the civilizations of ancient Sumeria and Egypt was largely the outgrowth of practical needs, either to foretell seasonal conditions affecting agriculture and pastoral pursuits or for the purposes of demarcating fields and building palaces and pyramids. In ancient Greece science emerged in the sixth century B.C. from mythology and religion in the Milesian philosophers' search for the universal nature (physis) of things, the stuff of which everything is made. This physis, however, was no merely observed substance, but was, as F. M. Cornford maintains, a metaphysical concept identified with soul and God. Thales thought that it was water, and he also believed that everything was besouled and that "all things are full of gods." His successor, Anaximander, believed that it was an indefinite blend from which a universal vibrant motion separated out "the Opposites": hot, cold, wet, and dry. These again combined to give fire (hot and dry), air (hot and wet), water (wet and cold), and earth (cold and dry). His pupil, Anaximenes, considered that the original substance must be air, the expansion and contraction of which would account for the diverse consistency and constitution of things.

These speculations were clearly attempts to understand permanence and change, and succeeding philosophers emphasized each of these features of the real while denying the other. Parmenides declared that what is not could not be, and thus only Being existed, and there could be no change, while Heracleitus asserted that everything was in flux and that nothing was permanent. Pythagoras then propounded the doctrine that all things were numbers, establishing the sciences of geometry and arithmetic in an attempt to account mathematically for sameness and difference. Empedocles alleged a constant alternation between the One of Parmenides and the flux of Heracleitus, in which the elements resulting from the combination of Anaximander's opposites, separated out from the all-inclusive sphere of the One, under the influence of Strife, and then combined again in varying shapes and proportions under the influence of Love. Anaxagoras advocated the hypothesis that things were constituted of minute seeds of innumerable different kinds, variously mixed under the direction of Mind (*nous*). Then Leucippus and Democritus, using the notions proposed by these earlier philosophers, put forward the theory that nothing existed except atoms (each having the permanent characteristics of Parmenides' Being) and nothingness, the void. Commonly experienced things were thus diverse combinations of atoms, not very different from Anaxagoras' seeds, and it is only "by convention" (in Democritus's words) that their qualities exist.

All these ideas were used and developed by Plato and Aristotle to construct a more inclusive and elaborate system, bringing to maturity the main concepts of Greek philosophy and science, between which no sharp line can be drawn. The intellectual structure that emerged included logic, psychology, ethics, and political philosophy, along with biology, physics, and metaphysics, the principles of which governed all the rest. The efforts of the scholars in the Academy and the Lyceum to account for the aberration of the planetary orbits gave rise to complicated theories of the composition of the heavens as crystal spheres bearing the fixed stars, the sun and moon, and the wandering satellites. So began the modern sciences of mathematics, astronomy, physics, and (with Aristotle's research) biology and psychology; but none of them were originally distinct from philosophy, with which they continued to be identified well into the modern era.

Not only did philosophy and science grow from roots in mythology and religion, but, as the philosophers expounded their different theories, these had repercussions upon religion and practical morality, as well as upon politics and law. This became noticeable with the influence of Stoicism and Epicureanism in the Hellenistic and Roman periods, modifying religious belief and shaping Roman law in significant ways.

The "nature" originally conceived by the early Ionian thinkers and which was not distinguished from soul or gods included the human soul and human nature, whose natural character was to think rationally, being the activity of mind (*nous*). For Aristotle nature (physis) was the primary matter or potentiality, the actualization of which was its form, and the ultimate form of all forms was God (Prime Mover of the heavens and the source of all physical motion), whose natural and appropriate activity was "the thought that thinks itself (*nóesis noeseòs*)." From these concepts, as they were adopted and modified by the Stoics, came the idea of the Law of Nature, which was at the same time the law of reason and of God, and which underlay both morals and the civil law, especially that branch of Roman law which applied to all nations universally.

Western science in its origins, then, cannot be divorced from religion, morality, politics, law, or social practice, and was dependent upon—was part and parcel of—a system of education (*paedeia*) inseparable from the whole pattern of Greek civilization. In their later developments, these interdependencies remained unbroken, so that what pertains to science, as such, is related to and has inescapable effects on all other aspects of social organization.

Science, we may conclude, is not a self-contained or self-sustaining activity. It is conducted only in a community that has reached a certain level of intellectual development, which involves and implies social organization, culture, art, and religion, as well as philosophy. While science is part of such a culture and cannot exist apart from it, these cultural activities, in their turn, are affected by the discoveries and pronouncements of the scientists. The "paradigm," therefore, affects all features of the culture and is intrinsic to them all in a radical manner.

The paradigm of ancient Greek science was the conception of the cosmos as a living being with a world soul of which the souls of gods and animals were specific manifestations. The Aristotelian paradigm developed in Greece over a period lasting from the sixth to the third centuries B.C., and it determined the form and practice not only of science, but also of religion and politics throughout Western Europe for the next millennium. Plato, developing the Pythagorean notion of eidos (the form of a number), taught that the material temporal world was "a moving image of eternity," that is to say, an imperfect copy of a range of perfect and immutable Forms culminating in the Form (or Idea) of the Good. The world, he thought, was a living creature with its own soul, the soul was akin to the Forms (and therefore immortal) and was the principle of self-movement. The material world and its soul had been created by the Demiourgos. But the Platonic Demiurge, who (in the *Timeaus*) constructs the world, is not an omnipotent God creating the universe out of nothing but simply one of the Pantheon who takes over the primitive chaos and molds it into structures modeled on the eternal Forms.

Aristotle also considered form as determining the essential nature of things, but in a progressive series, in which a presumed prime matter, having received a form (one of the four opposites—moist, dry, hot, and cold—constituting the elements), the matter thus formed became further matter to a higher form, and so on in ascending gradations, until the soul is seen as the form of living matter. Soul itself includes subsequent degrees of matter and form: the nutritive and reproductive soul (of plants), which becomes the proximate matter of the sensitive and appetitive soul (of animals), and that again is proximate matter to the rational soul (of humans). The supreme human activity is contemplation mirroring that of God, who is pure form—the Form of all forms—the thought which thinks itself, whose activity all lesser things imitate.

Living movement was directly attributed to the souls of animals, *animus* being the Latin form of the word for a soul. But human beings are capable not only of voluntary locomotion but also of the closer imitation of God, namely, thinking. God is thus the Prime Mover, the imitation of whose activity by the ethereal heavens is circular motion, the most perfect form of motion and the nearest to self-thinking thought possible for material bodies. Sublunar matter, if violently removed from its natural place, returns to it by "natural motion"—earth to the center, water next above earth, air above water, and fire beyond air. Rather than to Plato's Demiourgos, Aristotle's God is nearer to the Idea of the Good, which makes all things intelligible, all minds intelligent, and is the creating and sustaining cause of everything.

This paradigm, as we now see, had a long history and a long development, beginning as a body of mythology underpinning religion and morality, and in due course becoming philosophical and scientific. From the beginning physics and metaphysics were one and the same, growing out of Thales's search for the original stuff of which all things are made and developing from the doctrines of the Ionians through those of Parmenides, Heracleitus, and Pythagoras to the theories of Empedocles, Leucippus, and Democritus. Plato and Aristotle brought the conceptual scheme to full fruition, combining and refining earlier ideas and molding them all into one coherent system.

This paradigm permeated all Greek thought. In its mature form it was passed on to the Romans via the Stoics and Epicureans, and then it was transmitted to the mediaevals, first through the neo-Platonists and later through Arabian and Jewish philosophers (such as Al Farabi, Avicenna, Averroes, and Maimonides). Thomas Aquinas established it as integral to the teaching of the Catholic Church, but it was not much altered by the supersession of Christianity, with the Judeo–Christian God replacing Plato's world soul (coupled with the Idea of the Good) and the Prime Mover of Aristotle.

Throughout the Middle Ages the scientific conceptions remained much the same, especially with respect to motion. The earth was still regarded as the center of the universe with the planets and heavenly bodies, as Aristotle and his followers had imagined, embedded in crystal spheres, revolving round it in circular orbits. In their attempts to explain the aberrant movements of the planets, the pupils of Plato and Aristotle had devised ingenious hypotheses of crystal spheres revolving one within another, for some thinkers, on different axes, for others, in different directions, the eventual product of which was Ptolemy's device of inserting epicycles rolling between the spheres to account for apparent loops in the planetary orbits. Thinkers in the Middle Ages retained these ideas, believing that God kept the heavenly bodies in motion, and that all sublunar movement was the result of some form of propulsion originating from the motion of the heavens, if it was not produced by the free action of intelligent beings.

It is thus apparent that the world view that dominated Greek and Hellenistic civilization was not simply a scientific concept but was a general cast of thought permeating the entire outlook of the culture of the Mediterranean region, and one that persisted throughout Europe, with only partial modification by the spread of Christianity, until the Renaissance.

The same is true of the Newtonian conceptual scheme. It was born as a scientific revolution, but it transformed the mental cast of Western civilization, in all its ramifications, in far-reaching ways that are examined in Chapter 2.

Both Collingwood and Kuhn recognized that, as knowledge develops, the prevailing paradigm (or constellation of absolute presuppositions) comes under strain. In the course of what Kuhn calls "normal science," when scientists are exploring evidence that may solve the puzzles they have encountered in the application of their theories, anomalies and contradictions arise. At first, if these do not too seriously interfere with the progress of research, they may be ignored, or ad hoc explanations may be offered to account for them. But as they accumulate, they come to obstruct and paralyze the advance of knowledge, producing a crisis which is only removed when eventually a revolution occurs in the thinking and practice of scientists with the introduction of a new conceptual scheme. Collingwood likewise maintained that the constellation of absolute presuppositions changed when "strains" (the source of which he did not identify) occurred among them, so that they were no longer "consuponible." Modifications are needed for coherent advance, but Collingwood does not tell us how they come about. As it is the task of metaphysics, however, not only to reveal these presuppositions, but also to trace the changes and their causes, philosophy is necessarily involved in the process.

Kuhn maintained that scientific revolutions produced a change in paradigms which, as explanatory schemes, were incommensurable. One cannot, he believes, judge the theories of a former pardigm in terms of its successor. There is no "goal"— no truth—toward which science advances, and its progress is measured only by the increase it affords in the understanding of nature. Collingwood, we have seen, explained change in the constellation of absolute presuppositions as due to strains, the source of which he left unexplained, although he denied that they arose from incoherence (because absolute presuppositions, he averred, did not imply one another).

In fact, the revolutionary changes are invariably prompted by contradictions which occur when the previous concepts are applied to particular phenomena, and it is in the attempts to remove these contradictions that the changes are made. Pace Collingwood, coherence and consistency are the persistent aim of science, which for it, as for all else, is the criterion of truth. And, contrary to Kuhn's belief and the assertions of several other philosophers of science (e.g., Karl Popper), there is a logic of discovery as well as of confirmation. It is the dialectical logic (concerning the nature of which we shall say more in a later chapter) that generates contradictions and effects their removal (their sublation) in subsequent reconciliation of opposites—but of all this more anon.

Accordingly, the revolutionary changes, although dramatic, are not sudden, but occur in stages. Contradictions arise when current theories are applied to recalcitrant phenomena (as when attempts were made to explain the flight of an arrow by the Aristotelian theory of motion). Modification is then introduced into the conceptual scheme to preserve consistency, and when further difficulties arise, the entire scheme is replaced to provide a more coherent world picture. The new theories are not just conjectures (as Popper alleged), but are modifications of the old, which by stages introduce the new paradigm. The process takes place through the contributions of several important thinkers who make successive modifications to existing hypotheses, until one major genius sees how all these contributions can be fitted together into a more coherent system.

For example, contradictions that arose when thinkers tried to explain the flight of an arrow and other such terrestrial missiles in terms of the Aristotelian theory occasioned tensions in the conceptual scheme. Also, continuing difficulties in accounting for the persistently aberrant movement of the planets prompted the postulation of increasing complications as each new modification to correct inaccuracies required the insertion of more and more epicycles. The consequent complexity of tracing out the Church calendar and fixing the dates of religious festivals and observances resulted in yet further strains.

The first of these difficulties prompted the ingenious theory of impetus, a step toward the concept of inertia. Then the cumbersomeness of the Ptolemaic astronomical system led Copernicus at last to attempt to rationalize the structure of the solar system by proposing the hypothesis that the sun was at its center, and the earth, like other planets, revolved around it—so that (as he claimed in his prefatory letter to Pope Paul III) the system became more coherent.[4] This introduction of heliocentrism (it was in fact a reintroduction of the hypothesis put forward by Aristarchus of Samos) produced a radical revolution in the thinking of subsequent generations and initiated the new paradigm.

Impressed by the excessive and confusing complications of the Ptolemaic system of astronomy, Copernicus did not invent his new hypothesis—it was no mere guess. He revived Aristarcus's hypothesis that the sun was center of the planetary system and not the earth. Nor did he altogether abandon Ptolomaic ideas, but still made use of epicycles and eccentrics. He was followed by Tycho Brahe, who tried to develop an intermediary system with the earth at the center but with all the other planets orbiting the sun. Making use of Tycho Brahe's copious observations, Kepler (who was for a time Brahe's assistant) discovered that the orbit of Mars was an ellipse, and devised laws of planetary motion to account for this; meanwhile, Galileo was developing laws that govern the motion of falling bodies. Newton's genius was required to put all these advances together into a single coherent system. As he confessed, if he had seen further than others it was because he stood on the shoulders of giants. The procedure is one continuous effort to construct a coherent explanatory system embracing all known phenomena, and its result is a characteristic world view.

The advent of the Copernican hypothesis altogether subverted the previous *Weltanschauung*. If the sun is the center of a celestial system with the earth and the planets rotating around it, the earth becomes a celestial body, and Aristotle's distinction between the sublunar sphere and the outer heavens, crucial to his entire system, disappears. If day and night result from the earth's revolution on its own axis, while the fixed stars remain stationary, there is no longer any call for crystal spheres and no place for a Prime Mover; nor, in the mechanical conception of the physical world, which the Newtonian theory of gravitation (the final consolidation of the Copernican theory) introduced, was there any need for a soul (either of the world as a whole or of individual animals) as the source of self-movement.

Furthermore, the Aristotelian doctrine of the human soul and the whole structure of his ethics was based on the contention that sublunar bodies and minds, in their various degrees, imitated the activity of God. Once the Copernican conceptual scheme was admitted, this account of the movement of the heavenly bodies had to be abandoned, and God was excluded from the material world—still regarded as its creator, indeed, and the author of the laws of its movements, but apart and discontinuous from the mechanical system that those laws activated. The ethical teachings of Plato and Aristotle gave way to the Mosaic and Christian, but, as these too were alien to the forthcoming Newtonian paradigm, serious divergences arose, with deep-seated social and political consequences, which we shall presently consider. The revolution was not just scientific; it affected the entire conceptual scheme of the ensuing phase of European civilization.

Not only were astronomy and mechanics revolutionized, but with them philosophy and religion. The Aristotelian system that had been adopted by the Catholic Church with the teaching of Thomas Aquinas was rejected by the Humanists and Baconians of the new era in favor of the atomistic doctrines of Democritus and Epicurus. The conception of the Judeo-Christian God, which had developed in definite ways even before this scientific revolution, now had to be modified still further. Originally a tribal deity, Jahwe, with the later Hebrew prophets and Christianity, was reconceived as a universal, omnipotent Being, creator of and ruler over all His works. This notion had been combined with the Aristotelian postulation of the Prime Mover; but now a radical change was necessary. God was no longer a factor in the physical system. Initially He came to be conceived as a purely spiritual Being, the supreme Architect of a mechanistic universe moving according to eternal laws of Nature that He had decreed.[5] But it soon came to be felt that even this assumption was superfluous, and the effect on religious belief was profound. The Newtonian paradigm gave rise almost immediately to a conflict between science and religion, which has persisted in one form or another to the present day.

The divinely decreed law of nature had also come to be regarded as the basic criterion for social and political law, and, as it was equated with the law of reason, the outlook determining human social relations was reconstructed afresh. The new scientific paradigm affected the entire conception of human society and morals, along with religion and political authority. We shall presently examine in more detail what these effects have been.

The establishment of a paradigm, such as either of the two which dominated Western civilization from its inception in the eastern Mediterranean until the end of the nineteenth century, is never a sudden or immediate transition. In either case it took centuries. In ancient Greece (if we ignore the influence of ancient Egypt), the emergence of a scientific outlook began with Thales in the sixth century B.C., but the paradigm was not fully established until three centuries later with the work of Plato and Aristotle; nor was its effect on moral and religious practice fully felt until well into the Hellenistic era. At the time of the Renaissance, Copernicus conceived his hypothesis in the earlier part of the sixteenth century, but it was not until late in the seventeenth that Newton brought it to maturity with the publication of his *Principia Mathematica*. It took another century for the full effects of the new paradigm to make themselves felt with the Enlightenment, in the spheres of politics and religion.

Not only does a new paradigm take time to establish itself, a revolution in scientific ideas and in *Weltanschauung* is never immediate. It does not occur like a bolt from the blue, but is always a development from and modification of previous concepts. As has been said, Copernicus did not think up his heliocentric system ab initio; he found it in Archimedes' report of the theory of Aristarchus of Samos. Even earlier the Pythagoreans had entertained the view that the earth and the heavenly bodies revolved round a Central Fire. Nor was Copernicus's system altogether free of Ptolemaic features: He still made use of epicycles. Galileo grafted his new conceptions of inertia and momentum on to the theory of impetus that Buridan had elaborated in the fourteenth century. Further, the new paradigm included the belief in the particulate structure of matter, adopting the idea from the ancient theories of Leucippus and Democritus, gravitating bodies being conceived as mass-points. The same continuity in significant respects can be traced between the developments of electromagnetic theory in the nineteenth century and the new and revolutionary ideas of Einstein and the quantum theorists of the twentieth.

Anomalies and contradictions caused serious difficulties for scientists working with the Newtonian paradigm in the declining years of the nineteenth century. With Max Planck's discovery of the quantum of action and Einstein's theory of relativity, which removed some of the contradictions plaguing the prevailing conceptual scheme, a new paradigm came to be established which again transformed the conception of the nature of the universe. The effects for physics and physical cosmology have been dramatic and have invaded the biological sciences. But in other quarters the impact of the new paradigm has not yet been felt. The consequences of Newtonian thinking have persisted in morality and social relations, in economics and politics, while in philosophy there has even been a throwback to eighteenth century modes of thought, just when the influence of the new physics (in the metaphysical systems of such thinkers as Samuel Alexander and Alfred North Whitehead) was beginning to take hold. This reaction is a curious historical anomaly to which no clear parallel can be detected at the times of past scientific revolutions, even though, as has been observed, new paradigms take a long time to penetrate the thinking and practice of society in all its aspects.

Today, however, at the end of the twentieth century, the world situation differs drastically from what it was in the seventeenth. In the past the religions and philosophies of the Far East remained intact, in large part uninfluenced by European culture;

in fact, the influence, if any, went in the opposite direction. Eastern thought, on the whole, tended to be more mystical and holistic than Western, and outside Western Europe the initial effects of Newtonian science were negligible. Moreover, apart from that fact, the prevalence in its day of Aristotelian science and its accompaniments in philosophy and religion were never such as to threaten the welfare of the human race and its civilization, so that no dire consequences attended the fact that the Copernican revolution took two centuries to permeate the culture of Western Europe. The influence of Newtonian science has been much more far reaching.

Roughly at the same time as the Copernican revolution, and largely stimulated by the new understanding that the world was round, European exploration of the seas and other continents spread the influence of European culture far and wide. At a later date European colonization throughout the western and southern hemispheres, as well as in India and China, had a similar effect. The Newtonian paradigm meanwhile was so fruitful that science advanced spectacularly during the next two centuries, enabling an unprecedented development of technology that brought about the Industrial Revolution and mechanical inventions which vastly increased the capacity for all kinds of production and revolutionized the means of transportation and communication.

In consequence, European culture has, in the intervening centuries, girdled the earth and come to dominate the way of life of the less mechanized nations of the Americas, of Asia, of Africa, and of Australasia. European ways of thinking, European education, European art, and European philosophy have invaded the contrasting customs of other regions; but even more important, European technology and economic organization have overrun the entire globe. All of these aspects of social order bear the marks of Renaissance science and the habits of thought engendered by the Newtonian paradigm, so that now its effects are worldwide, and they have had vicious consequences which, at the present time, threaten the continuance of human civilization altogether, both by the devastations of world war and by disrupting the entire planetary environment, with the direst menaces to the future survival of all living things, embracing the human species and everything that it has achieved in the past.

This dismal prognosis however, is not inevitable. The new twentieth-century paradigm introduces a conception of nature as a single indivisible whole—a holism, the disregard of which exacerbates the destructive tendencies of present human practices based on contemporary technology. The wholehearted adoption of this new conception of nature and the instillation of an holistic habit of thought in ethical, social, economic, and political practice would counteract the deleterious effects of modern technical developments. Such penetration of the new ideas into the current way of life would, in the normal course of history, be gradual and very slow, but the present destruction of the planetary ecology is proceeding so rapidly that, if it is not promptly checked, it will (if it has not already) become irreversible. If human civilization and human life itself are to be saved from extinction in the foreseeable future, we cannot wait for the normal historical processes to take their slow and virtually imperceptible course. The means of salvation are already in our hands, if we could but recognize them, and somehow the infection of the new ideas inherent in the twentieth-century scientific paradigm must be encouraged to spread and its penetration of modern (or postmodern) habits of thought must be accelerated if our culture is to survive.

This is the theme that I shall endeavor to elaborate in the following chapters, examining first the implications and their consequences of the Newtonian paradigm and the way they have persisted beyond due time up to the present, precipitating our contemporary unprecedented crisis. Then I shall try to show how implications of the Einsteinian and Heisenbergian revolution reverse the disruptive tendencies of the previous conceptual scheme. How to bring about a general and widespread recognition of these implications and the corresponding revolution in outlook that they demand, apart from merely pointing them out and underlining them, it is difficult to suggest. What does seem clearly apparent is that unless such widespread recognition and the changes it requires in our habits of thinking occur in the immediate future, the long-term prospect for humanity will be bleak indeed.

NOTES

1. T. S. Kuhn, *The Structure of Scientific Revolutions* (Chicago: University of Chicago Press, 1962–1964), p. 10.

2. R. G. Collingwood, *An Essay on Metaphysics* (Oxford: Clarendon Press, 1940), p. 66.

3. See Giorgio de Santillana, *The Origins of Scientific Thought* (New York: Mentor, 1961); F. M. Cornford, *From Religion to Philosophy: A Study in the Origins of Western Speculation* (New York: Harper Torchbooks, 1957) (especially Chapter 5); Bruno Snell, *The Discovery of the Mind* (New York: Harper Torchbooks, 1960), Chapter 10.

4. Cf. E. E. Harris, *Hypothesis and Perception* (London: Allen and Unwin, 1970; reprint, Atlantic Highlands, N.J.: Humanities Press, 1996), p. 88.

5. Cf. R. Descartes, *Discourse on Method*, part 5, in *The Philosophical Works of Descartes*, vol. 1, ed. E. S. Haldane and G.R.T. Ross (Cambridge: Cambridge University Press, 1931), p. 109.

2

The Newtonian Paradigm

The world view that emerged from the Copernican revolution in the sixteenth century was of the physical universe as a vast machine. Galileo was the father of the science of mechanics proper. What the Greeks sought to explain was not so much motion as change in relation to permanence, and for them motion was simply a particular kind of change. For Aristotle motion was change of place, the alternation of potentiality and actuality. The theoretical idea of forces was entertained, if at all, only very dimly and late in the development of Greek science, being simply the awareness of the effects of the pressure exerted by the elements (air and water; in the case of earth, impact) as it became evident in the construction of useful instruments (such as the water clock). Mechanics as a mathematical science did not exist for the ancients.

After Copernicus, however, astronomers such as Kepler sought to define mathematically the orbits of the planets, and Galileo experimented by rolling balls down inclined planes with the aim of discovering mathematical laws governing the motion of falling bodies and flying missiles. Newton, building on their work, evolved a science of mechanics that applied equally to motion on earth and in the heavens. The whole physical universe was then conceived as one all-inclusive machine.

The primary character of the Newtonian view of the physical world is mechanism, and the exclusive constituents of a mechanical world are matter and motion. Newtonianism, therefore is essentially materialistic and mechanistic. For seventeenth century thinkers the objective world consisted of a system of material parts moved according to fixed laws by specific forces. The machine was external to the mind, the object of observation, and what constituted its objectivity was just this externality. Its intrinsic nature was its materiality and its mechanical construction—its mass and measurability.

Matter was distributed as bodies which moved through space in time at varying speeds under the influence of forces, some intrinsic (*vis insita*), others extrinsic (*vis impressa*). Of these the most important was gravity, intrinsic to all bodies, by which they attracted each other, with a magnitude proportional to the sum of their masses and the inverse square of the distance between them measured from their centers. To simplify calculation all bodies were conceived, in the last resort, as mass-points, and matter in general was conceived as particulate.

This mechanical picture carries with it three important corollaries. First, space was conceived as one all-encompassing receptacle, immobile and absolute, in which all bodies resided and moved, and time was a similarly absolute passage of successive events. Second, all matter, presumed to be particulate, was conceived atomically. Third, bodies (or particles) were thought of as mutually separate and independent, the relations between them being entirely external (their mutual gravitational attraction notwithstanding). A direct result of this attitude is to encourage reductionism as the appropriate method of explanation: the reduction of all complex phenomena to their simplest terms, and the examination of these first in their separation and then in the relations subsisting between them.

Knowledge about the mechanical universe, to be reliable, had to be entirely dispassionate, uninfluenced by any subjective feeling. The outer world was to be discovered by observation and measurement as accurate as possible and untrammelled by invention or wishful thinking. *"Hypotheses non fingo"* was Newton's motto. The observer was to survey the world, as it were, from the outside, as one perceives a view through a window, or the heavens through a telescope. Accordingly, there was no place in the external (objective) world for the observer, nor could its mechanical laws account for consciousness or the occurrence of mental phenomena. Consequently, an unbridgeable gulf divorced matter from mind, and the relation between them remained an impenetrable mystery. A concomitant requirement demanded of all scientific knowledge came to be that it should be value free, for values (other than quantitative) are not phenomena and cannot be observed. They are subjective preferences formed by the mind and merely projected onto its object. Fact and value were separate and incommensurable.

This separation of matter from mind is implicit in the distinction made by Galileo between primary and secondary qualities. Only what is measurable and quantitatively calculable he regarded as actually pertaining to external bodies. Accordingly, the language in which science is written, he maintained in *Il Saggiatore*, is mathematics; and all qualitative distinctions were held to be simply the effects on our sense organs of external bodies.

Another consequence of the mechanistic outlook is the opposition, emphatically expressed in the period of transition, to teleological explanation. Bacon satirically likened teleology to a vestal virgin, who could produce no offspring. Explanation had to be sought through efficient causation determined by the laws of mechanics, the assumption of final causes being no more than the empty assertion of tendencies to produce what was already known to occur, for example, in Moliére's farcical example, opium produces sleep because it has a dormitive tendency.

In summary, the characteristics of the Newtonian paradigm are

1. An absolute frame of space and time.
2. Materialism and mechanism.
3. Atomism.
4. Reductionism.

5. The assumption that all relations are external.

6. The demand for unbiassed observation and value-free science.

7. Rejection of teleological explanation.

8. Complete matter–mind dichotomy.

METAPHYSICS AND EPISTEMOLOGY

The effect of this scientific outlook on philosophy was incisive. Scientists declared that the best and only reliable method of investigation was by observation and experiment, so the questions immediately arose: How and why was this the case? How does observation record the facts? How does the external world come to be represented in the mind? How do we know that what we perceive is true of the external world? Now for the first time, the problem of knowledge becomes explicit and insistent.

The answer to the question depends on the relation of the mind to the body, and that again depends upon the relation of matter to mind. Three possible options were open to speculation: to maintain a comprehensive dualism and, in the case of our own knowledge, to attribute our experience of the material world to the intervention and veracity of God; to reduce mind to matter and attempt to incorporate it into the all-embracing machine; or alternatively to reduce matter to mind and to declare all reality to be ideal. All three of these options find expression in the philosophy of the era, but none of them could provide a satisfactory solution of the attendant problems.

Descartes, beginning from a rejection of all opinions in any way susceptible to doubt, questioned our assurance that waking consciousness actually revealed an external world. How, he asked, can dreaming be reliably distinguished from waking? What is the source and the criterion of indubitable truth? His answer was that the knowledge of the thinker's own existence, while consciousness is occurring, cannot, without self-contradiction, be doubted or denied. The existence of the *ego* is the indubitable basis of knowledge from which all else is to be derived. Reflection upon the way in which he reached this unchallengeable starting point led him to the conclusions: (1) that the criterion of truth is clarity and distinctness, and (2) that the correct method is one of analysis of the complex into the simple ("simple natures") and of proceeding from one step to the next by indubitable (clear and distinct) intuition (*deductio*). The *ego*, however, is an isolated "I"; so in it, along with the postulation of simple natures, we have a reflection of the atomism and reductionism of the prevailing paradigm.

Descartes confessed to a confused if habitual belief in his mind of the resemblance to physical things of the sensible ideas he experienced, but he found sensuous appearances unreliable. How then can we be sure that our perception does represent to us material reality? In the end, he appealed to the veracity of God to provide assurance of the truth of our knowledge of the external world, ideas of which are not always of our own making and must therefore be produced by some other "faculty" than our own. He derives the certainty of God's existence from the certainty of his own, for he knows himself to be an imperfect and finite creature, which necessarily implies that he has in his mind the idea of a perfect being (from which his idea of imperfection must be derived), and such an idea of perfection must, of necessity, include existence. Moreover, as he is himself finite and defective, he could not be the cause of such an idea of perfection, the "objective essence" of which (that is, its ideal nature as presented to the mind) exceeds the "formal and eminent essence" of his own *ego*. His idea of God therefore could be produced only by a being "formally and eminently" adequate to cause the "objective" perfection of the idea—namely God Himself.

The cleavage between mind and matter is thus reaffirmed by Descartes, first in his initial evacuating doubt of the existence of external bodies (even including his own), which is not immediately dispelled by the pronouncement *cogito ergo sum*, which establishes the existence of his conscious mind. In Descartes's words, "[The] notion of thought precedes that of all corporeal things and is the most certain; since we still doubt whether there are any things in the world, while we already perceive that we think."[1]

Second, he can be assured of the truth of perceptive knowledge of the world only by his confidence in the veracity of God, whose perfection could not admit of His allowing us to be deceived by what we clearly and distinctly perceive. Where material things are concerned, this extends only to their mathematical properties (their primary qualities). He then declares Thought and Extension to be utterly separate and different substances, both created by God, but otherwise independent and self-subsistent, the relation between which in ourselves is obscurely adumbrated in dubious speculations about the function of the pineal gland.

What is to be noticed here is the separation, not only of matter from mind, but also of God from the created world. The mechanical universe is God's creation, as human machines are the work of engineers and mechanics. God is on one hand the divine architect and engineer, on the other the creator of the human mind to whom He gives the light of reason. The human body, like all other bodies, including those of all other animals, Descartes considered to be a mere automaton, on which (unlike those of the beasts) the free rational will acts and the action of which the soul directs via mysterious interaction with the pineal gland.

Those who followed Descartes strove resolutely to overcome the dualism. Spinoza did so by reducing Thought and Extension to attributes of God in whose substance their finite modes were identical. Leibniz reduced extension to a confused idea of the relations between atomic ("windowless") monads, themselves immaterial souls. Spinoza, rightly or wrongly, has been considered a materialist and is professedly a determinist. In fact, he is neither in the sense often attributed to him. Although the influence of the Newtonian paradigm on his thought is significant, he succeeded in large measure in transcending it. Leibniz tended more obviously to the idealist pole of the dichotomy, exerting important influence on later thinkers, while still reflecting the atomistic tendency of the age. He too was able to transcend the paradigm to some extent, for he was keenly aware of the metaphysical difficulties it occasioned and sought to overcome them by reverting to final causes and substantial forms. We shall see anon that these philosophical moves anticipated much later conceptual developments.

The British Empiricists, however, express most typically the presuppositions of Renaissance science. Hobbes explicitly and emphatically adopts the materialist option and attempts to reduce all mentality to matter and motion:

The universe, that is, the whole mass of all things that are, is corporeal, that is to say, body, and hath dimensions of magnitude, namely length, breadth, and depth; also every part of body is likewise body, and hath the like dimensions, and consequently every part of the universe is body, and that which is not body is no part of the universe.[2]

Here we have not only the assertion of universal materialism but also, by implication, the reality solely of primary qualities—dimensions of magnitude. Sensible qualities, we are told, are but "fancy":

All which qualities called Sensible, are in the object that causeth them, but so many several motions of the matter, by which it presseth our organs diversly. Neither in us are they anything else, but diverse motions; (for motion produceth nothing but motion). But their appearance to us is Fancy, the same waking, that dreaming.[3]

Hobbes makes no effort to explain the nature of "Fancy," nor how mere motion can "appear," or to what.

The epistemological problem is more evident in John Locke's account. He adopts the current presuppositions and develops their consequences only to become embroiled in problems that, on his premises, are insoluble. The premises are what the scientific paradigm prescribes, and Locke's conclusions from them reveal inconsistencies of which he seems unaware, but which his successors strive heroically to remove.

Although Locke professes in his introduction to *An Essay Concerning Human Understanding* to refrain from meddling with physical considerations of the mind, he follows Hobbes in considering it "evident that some motion must be thence [from external objects] continued by our nerves or animal spirits, by some part of our bodies, to the brains or the seat of sensation, there to produce in our minds the ideas we have of them."[4] Yet again no attempt is made to explain how this physical motion can be converted into idea. The gulf between body and mind remains impassable.

Atomism is adumbrated in Locke's insistence that simple ideas (either of sensation or reflection) are the original source and first beginnings of all knowledge, an atomism that becomes fully explicit in Hume. Simple ideas, Locke says, are the results of powers in the object to affect our senses. These powers are what we call the qualities of the affecting bodies. In the things themselves they are so united that there is no separation between them, but the ideas they cause in us are mutually distinct, separate, simple, and unmixed (a contention that again has profound consequences in Hume's development of the doctrine).

These ideas we acquire from experience, prior to which the mind is assumed to be like blank paper "void of all characters." The mind, Locke avers, before it receives simple ideas of sensation, is like an empty cabinet or a dark room, into which knowledge enters through the senses as light through a window. Sensation is thus the original and only source of knowledge of the external world; and, as finally turns out, is the sole test of its truth about the real. So the object remains external to the mind, and the knowing subject is no part of it.

Primary qualities, Locke tells us, are those which are inseparable from the body they qualify: solidity, extension, figure, motion or rest, and number, and of them our ideas are (he maintains) exact copies—although how he, or anybody else, could discover this fact, he does not reveal. The primary qualities depend on the real internal nature of the body, which is intrinsic to its imperceptible particles. Locke is thus committed to the belief in the atomic, particulate structure of matter, of which the primary characters are measurable and quantifiable. All other qualities, such as heat and cold, color, and sound, taste, and odor, are secondary. They are effects, he says, of imperceptible particles impinging upon our sense organs, and there is nothing in the bodies themselves resembling them. They are purely mind-dependent.

Starting from this position Locke cannot but conclude, first, that we can know only our own ideas, and, second, that their truth consists in agreement between them. But he promptly contradicts these conclusions by alleging that knowledge of real existence consists in the conformity between our ideas and the reality of things, which, he admits, we can know only by the intervention of the ideas. It is a conformity, therefore, knowledge of which, on his own showing, must be beyond our reach.

Berkeley and Hume, without abandoning Locke's starting point, attempt to develop his position in such a way as to avoid this contradiction. Berkeley does so by abandoning the external world altogether and confining all reality to ideas. But, as they are separate and unconnected one with another, he denies the possibility of their being abstract, although the idea of any particular thing can be used to stand for others of the same kind. He can then find no necessary connection between ideas, which, he maintains, are inert and cannot be causes. From Locke's beginnings there thus develops a psychological atomism that reaches its fruition in Hume's analysis. According to Berkeley our ideas are true (no mere "false imaginary glare") so far as they occur in regular order ("the settled order of nature" maintained by God). Here, once more, the deus ex machina is called in to provide the criterion of truth; but we have no access to God's mind, and so can recognize this "settled order" only by experience, which reveals no necessary connection between ideas. It is a criterion therefore only precariously available to us.

Hume, in accordance with the contemporary outlook, asserts that the only solid foundation for a science of man is experience and observation: "It is no astonishing reflection to consider, that the application of experimental philosophy to moral subjects should come after that to natural."[5]

For Locke's simple ideas, Hume substitutes simple impressions; ideas, he holds, are paler copies of these. They are variously associated in the course of experience and combined by imagination. Knowledge is divided into the intuition of relations between ideas and the perception of matters of fact. The latter can be acquired only through experience, and prediction depends on the inveterate belief that impressions constantly conjoined in the past will recur together in the future. This is the source of our belief in causal connection. That the future will resemble the past is the principle of induction, which is the only legitimate method of science, whose theories can never be demonstrated necessarily and are at best only probable. But, adhering to the empiricist presumption that all knowledge is derived from sense impressions, Hume demonstrates incontrovertably that inductive inference can be justified neither by reason nor by experience. He argues that there is no sense impression of necessary connection between cause and effect and concludes that the idea originates from the strong feeling of expectation, which enhances the idea of the expected consequent whenever the impression occurs that has hitherto preceded it. As there is no sense impression, so there can be no idea of necessary connection and accordingly no universal ideas. The best we can attain to is empirical generality, stemming from the belief that the future will resemble the past; but this we cannot derive from experience (for obvious reasons) and, as to deny it is not to contradict oneself, it cannot be established by reason.

Just as Hume can find no impression of necessary connection, so he finds none of the self, and therefore no idea of self. When he looks into himself, he declares, he discovers only a bundle of impressions and ideas occurring in rapid succession. Descartes's *ego* thus goes by the board, as do clearness and distinctness as criterion of truth; for this Hume substitutes the strength and vividness of the idea as the mark of belief.

As the sole source of knowledge of matters of fact is sense-impression, judgments of value can be no part of it. Fact and value for Hume are completely separate, and neither can be deduced from the other. What is can never determine what ought to be. That is a matter of preference and choice and is essentially subjective. We are not led to it by reason, which is, and should be, "the slave of the passions"—the instrument for calculating means to ends.

The disappearance of necessary connection between ideas undermines the conception of efficient causality. The idea of necessary causation is thus reduced to a habit of mind to associate ideas that have been constantly conjoined in past experience.

The elimination of necessary connection removes universality from our knowledge and with it any criterion of objectivity. Knowledge crumbles away into a congeries of impressions, the source of which is unknown, and their paler copies, ideas. Belief is said to be simply enhancement of the strength of ideas. Hence Hume reduces all reasoning (even demonstrative, in which we can never be sure of avoiding mistakes) to probability, all probability to belief, and all belief to a habit of mind produced by frequent association of ideas. The structure of the external world he shows to be purely the product of imagination, in the last resort condemned as illusion. The final result of Empiricism is thus total skepticism.

But for the present we are not concerned with this denouement; we need note only the adherence to atomism among impressions and the reductionism implied in the demand Hume makes that in order to validate an idea, one must identify the (atomic) impression of which it is a copy. The corollary consequent upon this reductionism is the externality of relations (lack of necessary connection). Hume declares that "whatever objects are distinguishable are separable," and "our distinct perceptions are distinct existences" between which there are no connections, only fortuitous association.[6] Relations between them, therefore, must be wholly external. The method of science is induction, to which in the final issue even mathematics is reduced; and objectivity is ultimately eliminated. To ask whether body exists or not, he tells us, is vain. Nature has determined us to believe that it does, but we can establish the fact by no sort of reasoning. Thus the external world is banished from the mind and ultimately demolished as illusion. The effect of skepticism on the practical conduct of life we shall consider later. It was not immediately felt and has taken two centuries fully to develop.

The skeptical conclusion of Hume's epistemology awakened Kant from what he called his dogmatic slumber; but the way in which he strove to remedy the situation was by finding the source of universality and objectivity in a priori synthesis performed by the transcendental subject and applying only to phenomena, not to things in themselves. Salient elements of

Cartesianism and Empiricism are thus preserved: the singular independence of the *ego* and the subjectivity of knowledge, on the one side, and the unreachable externality of the real, on the other. Kant, however, made a significant stride in the direction of what later proved to be the germ of a new paradigm. He recognized the necessary coherence of experience and the wholeness demanded by the a priori synthesis accomplished by a unitary subject, and he realized that this wholeness was the essence of organism and the basis of teleology in natural living forms. To the importance of this premonition we shall return, in due course.

MORALS AND POLITICS

The Middle Ages inherited from the Stoics the conception of Natural Law. For the ancients this was the law of reason, and for the mediaeval church, while retaining this connotation, it was also conceived as the law of God. Natural Law was held to govern all things, both the physical world and human relations; it was as much the moral law and the law governing society, dictated by reason, as the law governing the movement of the heavens. After Newton, however, the idea of a law of nature changed. It was a general statement of causal relations determining the motion of material particles and regulating the working of the celestial machine. Knowledge of these laws was knowledge of matters of fact, so they could not be moral laws, which prescribe what ought to be done. What is the case (it was held), cannot determine what ought to be the case. The laws of nature were still held to be what God had decreed at the creation, but, once established, they would keep the celestial clockwork running, and it would continue to function without further divine intervention, so that even this proviso came, as we shall presently notice, to be significantly modified. Nature was now the material objective cosmos undisturbed by human interference, and the natural condition of man was thought of as what prevailed prior to the institution of civil society, in which the law was prescriptive, not merely descriptive. The Aristotelian dictum that man is by nature a political animal was no longer accepted. By nature human beings were just animals of the forest.

At the same time, the Cartesian egoism led to the conception of the person as a separate and isolated individual, yet one nevertheless gifted with the light of reason, so that Natural Law, as the law of reason, was still held to regulate human conduct. Accordingly, individualism and enlightened self-interest became the dominant concepts prevailing in the thought of writers on moral, legal, political, and economic theory, as determining human behavior. By nature, man was entirely free from all restriction, and the only moral obligation that could or should be imposed upon the free exercise of his will was what was rationally necessary to preserve his life and to prevent some persons from limiting the freedom of others. Consequently individual liberty becomes the paramount objective, and the imposition of any other than natural limitation upon it can only be contractual and by mutual agreement.

For Hobbes, the natural man was a self-seeking, independent individual competing with all other animals for the means of survival. In the state of nature therefore men were, by nature, enemies. But as the condition of strife and hostility was intolerable and unproductive, reason dictated that men contracted to give up their natural freedom to exercise their individual powers to one man (or possibly a group), who assumed sovereign power over the rest, and established a polity maintaining law and order. The resulting civil society was thus still based on the law of nature understood as the law of reason. It was a sovereign state, self-determining and independent, retaining the original relationship to other sovereign states of men in the state of nature—that of war.

The same individualism prevails in the thought of Locke, although he entertains a different conception from Hobbes's of the state of nature. For Locke, it is an orderly and peaceful condition, in which individuals have inalienable rights, inherent in each of them as persons, prior to the establishment of the civil state. Their motive for contracting out of the state of nature is simply to escape the inconvenience of each person's having to defend these rights individually against the claims of others, without the advantage of "indifferent and upright judges." What I wish to stress is the atomic individualism that gives each person independent of every other an inherent circle of rights and irrespective of any social cooperation. These natural rights, according to Locke, limit the justifiable powers of any government—a limitation enshrined in the original contract presupposed as the basis of the civil state. The political consequence is a doctrine of limited sovereignty and a persistent tension between claims of the individual and the acts of government, against which the rights of the governed have to be defended. The device used for this purpose was set out by Montesquieu as a balance of powers between the three functions of political rule, the legislature, the executive, and the judiciary; and the persistent opposition of personal claims to government intervention was reflected in the works of such writers as Tom Paine, Henry D. Thoreau, John Stuart Mill, and Herbert Spencer. In practice, the result has been the adoption of the ideal of liberal democracy acclaiming individual rights against government action and fostering a constant hostility to and suspicion of politicians and government agencies. The extreme (if somewhat rare) form of this individualism is the advocation of complete anarchy and the total abolition of government, elevating individual rights at the expense of social order.

Concomitant with this idea of individual self-gratification, as it is found in Hobbes and Locke, is the notion that pleasure or happiness is the ultimate aim of all human beings, and some form of hedonism runs through all the writings of British Empiricism, culminating in the doctrine of Utilitarianism. It is reflected in the preamble of the Constitution of the United States, which affirms that all men are created equal and have inalienable rights to life, liberty, and the pursuit of happiness.

Morality thus becomes largely a matter of seeking personal advantage, and as pleasure is a subjective feeling and tastes differ, the strong implication is that values are purely relative to individual and social preferences. However, the prevalence, in modern civilization, of the influence of Judeo-Christian religion (with its exhortation to love God and your neighbor, the product of the previous world view) has occasioned a conflict between this hedonistic tendency and traditional belief, to circumvent which writers like J. S. Mill, went through self-refuting contortions, introducing qualitative criteria to modify that of the sum of pleasures. The effects on religion of the Newtonian paradigm, however, we have yet to investigate.

Another feature of the tendency in ethical theory to individualism and subjectivism is the prevalence, especially among the British moralists, of the doctrine of a moral sense, which enables every person to know intuitively what is right or wrong and submits everyone to an obligation to pursue the first and refrain from the second. Kantian ethics retains this characteristic identifying moral obligation as the categorical imperative of the rational will to preserve universality and self-consistency independent of the motive or consequences of the act. Moral sense is thus relative to the person, and even for Kant, universality, as we shall presently see, is confined to the noumenal will and does not belong to phenomenal motivation.

ECONOMICS

The emergence of economics as a separate science, pursuing a quasi-empirical method, was a natural outgrowth from the prevalent scientific paradigm. This new science incorporated the ideas that had surfaced in moral and political thinking: the pursuit of personal advantage and the inadvisability of government interference. The pursuit of individual advantage derives from natural appetites and the drive to supply natural needs; hence the laws governing demand and supply will be in large measure natural laws. Consonant with this position are the assumptions that the driving forces of all economic activity are demand, supply, and the pursuit of profit.

In the pioneering work of Adam Smith, father of modern economics, this habit of thinking is in evidence throughout. He explains in *The Wealth of Nations* that as standards of living would be minimal if each individual attempted to produce everything requisite for survival, in advanced societies wealth results from the division of labor and the consequent exchange of the produce surplus to the need of the producer. Thus, although the inevitable interdependence of persons is recognized, their mutual help is said to depend on the self-love of each:

Man has almost constant occasion for the help of his brethren, and it is in vain for him to expect it from their benevolence only. He will be more likely to prevail if he can interest their self-love in his favour, and show them that it is for their advantage to do for him what he requires of them.[7]

The resulting distribution of goods is based on the human propensity to "truck, barter and exchange." This again is rooted in the pursuit of enjoyment and advantage and is determined by the adjustment of supply to demand. The price of labor everywhere depends, says Smith, upon the contract usually made between the two parties, laborer and employer, "whose interests are by no means the same—the workmen desire to get as much, the masters to give as little as possible."[8] Likewise, the price of commodities is determined by "the higgling and bargaining of the market."[9]

Despite his recognition of the interdependence of the producers of different commodities, Adam Smith never suggests that the economic enterprise could be a joint, cooperative venture. The advantage of the individual is in every respect the determining economic factor, and this is said to result automatically in the advantage of the society. The individual can judge of his own advantage better than can any politician; so Smith argues against state interference and against restraints on foreign exchange to protect domestic industries, which force people to buy home produce at higher prices than they would have to pay for imported goods.[10] The argument for free trade is thus initiated and the operation of the free market as producing the greatest advantage to society (viewed as a collection of individuals) through the unrestrained pursuit of the advantage of each. The presumption is that the natural laws of the market economy, left to itself, will produce the desired result, just as the natural laws of the celestial mechanism will ensure its working, without interference from external influences.

This is the prevalent frame of mind of capitalism, which developed contemporaneously with Newtonian science, bringing about a dichotomy in society between employers and employed, creating the class distinction between bourgeoisie and proletariat, and progressively increasing the chasm between rich and poor. It was this discrepancy within capitalist society that exercised the mind of Karl Marx and stimulated the theoretical analysis which issued in the ideology of Communism. It is, further, significant that the Marxist doctrine is professedly materialistic, clearly evidencing the influence of the Newtonian paradigm.

RELIGION

For the ancient Greeks and the mediaevals, soul and spirituality pervaded the entire universe. Jews and Christians alike proclaimed,

> The Heavens declare the Glory of God.

The first effect of the Copernican revolution upon this relationship was to separate nature from God as its Supreme Architect. The world was still conceived as God's creation, but it was distinct from Him, and He was beyond and apart from His finite product. The created world became a machine operating automatically according to fixed laws which God had established; but this assumption was simply stipulated. It was a residue from the previous paradigm and was no part of the new scientific theory.

The first conflict between the new science and religious teaching manifested itself in the reversion from the authority of the Church to the independence of individual conscience which prompted the Reformation. This again is a reflection of the individualistic trend prompted by atomic thinking. The next was the condemnation by the Church of the writings of Galileo. As the celestial mechanics developed and the physical theory became more complete, the divine influence became redundant, and Laplace was able to assert that he had no need of that hypothesis. Atheism and agnosticism became more widespread and typified the "rationalism" of the Enlightenment, expressing itself in the work of such writers as Voltaire and Hume. The traditional proofs of the existence of God were shown by Kant to be invalid, and religion was relegated to the sphere of faith, altogether separate from and alien to reason and science. Segregation of revelation from rational understanding emerged in the writings of Bayle and Diderot and culminated in the thought of Kant, Jacobi, and Schleiermacher. On the one hand, this relegation of religious belief to faith reinforced the tendency to relativism in religion and morals (as Hegel was soon to point out); on the other hand, it was now an easy step to cast doubt upon faith and revelation and to use science as the general support of atheism.

In the thought of John Stuart Mill, Utilitarianism and religious disbelief go hand in hand, and the attempt of William Paley to stress the evidence of contrivance in the natural mechanism as a proof of God's existence was finally overthrown by Darwin's demonstration that all appearances of purpose could be accounted for in terms of chance variation and natural selection (a fresh resort to mechanism). Darwin himself confessed to a lack of all religious belief, and his friend and colleague, T. H. Huxley, adopted the term agnosticism to define his doubt concerning the Supreme Being.

Nietzsche summed up this tendency in his declaration that "God is dead" and based the assertion on the scientific instauration. In so doing, however, he became involved in self-contradiction, implicitly rejecting science when he indulges (in *Beyond Good and Evil*) in a diatribe against reason.[11]

The corrosion of religion was thus inherent in the scientific paradigm introduced by Copernicus, and its evacuative influence upon religious belief progressed throughout the following three centuries to emerge fully fledged in the writings of Feuerbach, Nietzsche, and Marx.

CONCLUSION

The Newtonian instauration ushered in a period of unprecedented scientific advance, not only in physics, but subsequently in chemistry, biology, and geology. A direct result was the Industrial Revolution, the multiplication of mechanical inventions, and the consequent increase in production and transportation. Combined with the influence of the newly discovered theory of evolution, these advances encouraged a belief in continuous progress, the conviction that intellectually, socially, and morally, things had always and would continue to get better and better. This idea was not counteracted by the actual historical facts that, in consequence of the insistence on individual rights, the unquestioning claim to sovereign independence by states, the tacit assumption of subjectivism in morality, and the amoral profit-seeking operation of economic forces, the period was marked by social crises, violent political revolutions, and wars of increasing scope and devastation. These upheavals were, in fact, rather heralded as major strides in the advance toward liberty, democracy, and higher standards of living.

Wars followed one after another. The appeal to individual conscience and the schisms in the Church following the Reformation led to the wars of religion in the sixteenth and seventeenth centuries. This and the claims to individual rights sparked off the English Civil War. The claims of national sovereigns, especially with respect to overseas colonies, gave rise to continual wars in the eighteenth century. The pursuit of individual liberty led to the American and the French Revolutions and to the Napoleonic wars at the turn of the nineteenth. Industrial mechanization and the accumulation of capital wealth were accompanied by the impoverishment of the working class, leading to labor unrest, trade-unionism, and Marxism. Colonization was accompanied by the exploitation of resources worldwide and of indigenous colonial populations. And the celebrated "march of progress" mainly took the form of the spread of European culture, molded as it had now been by Newtonian science, over the entire surface of the planet.

NOTES

1. R. Descartes, *Principles of Philosophy*, Part 1, viii, in *The Philosophical Works of Descartes*, vol. 1, ed. E. S. Haldane and G.R.T. Ross (Cambridge: Cambridge University Press, 1931), p. 221.

2. Thomas Hobbes, *Leviathan* (Oxford: Clarendon Press, 1943), p. 524.

3. Ibid., pp. 11ff.

4. John Locke, *An Essay Concerning Human Understanding*, ed. A. C. Fraser (Oxford: Clarendon Press, 1894), book 2, chapter 8, paragraph 12.

5. David Hume, *A Treatise of Human Nature*, ed. L. A. Selby-Bigge (Oxford: Clarendon Press, 1888), Introduction.

6. Ibid., book 1, part 1, section 7.

7. Ibid., book 1, chapter 2.

8. Adam Smith, *The Wealth of Nations* (New York: Random House, 1937), book 1, chapter 7.

9. Ibid., chapter 5.

10. Ibid., book 4, chapter 2.

11. E. E. Harris, *Atheism and Theism* (Atlantic Highlands, N.J.: Humanities Press, 1993), p. 4.

3

Twentieth-Century Residue

At the opening of the twentieth century, the Einsteinian revolution brought a new scientific paradigm; but its presuppositions have not yet penetrated such aspects of society as are not directly concerned with scientific research. These are still dominated by Newtonian habits of thinking. In philosophy this hangover is perhaps more apparent than elsewhere, but it is traceable also in ethics and moral practice, in politics and economics, and in the relations between religion and science.

PHILOSOPHY

In one stream of contemporary philosophy, there has been a reaction against Newtonianism, but it was not derived from the new twentieth-century scientific concepts, and its effect has been merely negative. Husserl became conscious of a crisis in European science, but his diagnosis was rather a throwback to Kant and Fichte, who had sought to counteract the epistemological difficulties of Empiricism and the ethical dereliction of

Determinism by grounding the objectivity of scientific theory on the spontaneous activity of the knowing subject, subordinating objectivity to a subjective act which claimed to be free. Here the influence of the Cartesian *ego* persists. The science that Husserl, and after him Heidegger, criticized is Newtonian science; but what they put in its place is no product of the new Einsteinian paradigm, but is what Husserl calls "the natural attitude" and what Heidegger designates "being-in-the-world"—the naive consciousness of common sense. Husserl makes objective knowledge relative to phenomenal consciousness, harboring a subjectivism (akin to Kant's and Fichte's) in which the seeds of skepticism are unintentionally watered and kept alive. The tap-root of this philosophical growth is still Descartes's transcendental *ego* (that likewise nourishes the efflorescences of Existentialism), which not only isolates the individual subject, from whom no bridge can be found to other minds, but also carries the infection of relativism. But relativism is the twin sibling of skepticism, for where there is no objective criterion there can be no truth, and where there is no truth, no assertion can be accepted, not even that which denies objectivity. As Spinoza warned us, the consistent skeptic must be dumb.

The taint of skepticism has become ever more apparent in the ramifying offspring of Phenomenalism: in the relativism inherent in Structuralism, in the reaction of Poststructuralism, and the wrecking demolition of Deconstructionism. Similar trends can be traced in Historicism, Hermeneutics, and Postmodernism. These developments are all infected with the viruses of seventeenth-century Cartesianism on the one hand, and the positivistic derivatives of Newtonianism on the other, the line of descent through Auguste Comte being plainly discernible in the works of Michel Foucault and Jaques Derrida.

The other stream of twentieth-century philosophy is a curious historical anomaly. Just when the first philosophical expressions of the new scientific paradigm were beginning to emerge in the works of Henri Bergson, Samuel Alexander, and Alfred North Whitehead, the main stream of current philosophy was diverted, returning to eighteenth-century Empiricism.

The Logical Positivism of the Vienna Circle returned to Hume's rejection of metaphysics, giving the same reason: that it was not empirically verifiable (contained neither observable matter-of-fact nor mathematical reasoning), and its successive off-springs have regenerated Hume's skepticism in the work of Richard Rorty.

The early writings of Wittgenstein reasserted, in company with Bertrand Russell, the atomism of the former era, both factual and logical. As Hume could find no necessary connection between impressions or ideas, so Wittgenstein declares facts to be atomic, asserting that the world divides into facts, any one of which can either be the case, or not be the case, and everything else remain the same. It follows that factual propositions (which are said to show forth the form of the facts) will likewise be logically independent and atomic. As a result there can be no genuine implication between statements of fact. What has been called "material" implication is no more than the fortuitous coincidence of the truth values of unconnected propositions. Logical implication was reduced to tautology, and all relations were conceived as external. The only acceptable logic was formal symbolic logic, which is purely extensional, applicable only to collections (sets) of bare particulars.[1]

The source of factual knowledge was held to be entirely observational, so that factual truth could be found only in the empirical sciences, and philosophy was confined to mathematical logic. The only accepted method of empirical science was induction, and, despite numerous efforts by such writers as Hans Reichenbach and Nelson Goodman, no valid justification of inductive inference could be found that was invulnerable to Hume's proof that none was possible. Bertrand Russell had to confess that the difficulty could be avoided only by abandoning Empiricism to the extent of accepting the principle of induction as an axiom or a priori rule.[2] Every other attempt to justify inductive reasoning on empiricist principles has fallen foul of Hume's strictures.

Thus, for contemporary Empiricism, metaphysics has been jettisoned, philosophy has been reduced to formal logic (which is tautological and admittedly can deliver no positive information), and empirical science, the only acknowledged source of factual knowledge, is still vulnerable to Hume's critique, depriving it of universality and reducing it to a mere habit of belief.

It followed that philosophy, so eviscerated, could offer no insight into moral, political, or any practical problems, and philosophers in the main repudiated all claim to offer advice or guidance. Philosophy became an esoteric discipline in which students lost interest and for which other pundits lost respect. The task that Collingwood had set for metaphysics was totally ignored, for in general disregard of the sciences now held to be strictly external to their province, and disowning any attempt to "pontificate" over scientific concepts, philosophers were no longer competent to carry it out.

ETHICS

Not only was metaphysics repudiated, but also ethics and political philosophy. Ethical statements, being observationally unverifiable, were held to be simply expressions of feeling (of approval or disapproval), or else were prescriptive, and were thus noncognitive. The proper fields for the study of morality and values therefore were psychology and sociology. Political theories were denigrated as ideologies, generated by class prejudices or partisan interests.

Throughout the twentieth century, ethical theory has remained under the spell of the Newtonian scientific outlook and has been conditioned by the Cartesian dichotomy, the rift between nature and mind, and the incommensurability of fact and value. This severance had infected Kantianism, which has been revived in recent years, not only by Nicholai Hartmann, but in other forms of deontic theory. The dualism appears in Kant's doctrine as that between phenomena and noumena, in terms of which he seeks to solve the problem of free will by confining causal determination to the former, while assigning the rational will to the latter. Consequently the autonomous moral will is devoid of empirical content, while human action, so far as it is phenomenally experienced, can only be judged heteronomous, and therefore (inasmuch as nothing is morally good except the good will) immoral. Moral objectivity, in consequence, is inapplicable to phenomenal behavior and the scientific categories that assure objectivity in knowledge are incompetent to prescribe moral laws. "Ought" remains underivable from "is." Despite Kant's intention, therefore, conduct as perceived phenomenally was determined by natural impulse, and practical precepts were still vulnerable to the taint of relativism, which in the present age has become prevalent.

Twentieth-century ethics has in the main been either deontic or utilitarian. Deontic theories such as were expounded by W. D. Ross and H. A. Pritchard declared moral duties to be absolute and their content to be self-evident. They declared that we recognize intuitively what is right and obligatory (e.g., the duty to fulfill promises). But it had ultimately to be conceded that intuitions differ from one person to another, and what is prohibited in one society is considered permissible in another; so relativism had not been circumvented.

Empiricist ethics has constantly been either overtly or tacitly hedonistic and utilitarian, assimilating moral judgement to subjective feeling and denying the derivability of fact from value. Moral statements have been classified by empiricists as either optative or prescriptive, expressing no more than preferences and feelings of approval or disapproval. They are thus purely subjective and relative to persons, or at best to the social unit. The general aim of action is taken to be some sort of utility, the ground of which is pleasure and/or subjective satisfaction.

A suppressed form of Utilitarianism can be detected even in the writing of a thinker such as Kurt Baier, who attempts to distance himself from hedonism.[3] He alleges that the primary "reason" for action is "enjoyment"; but what different people enjoy differs with temperament, so objectivity is still in jeopardy. The "moral reasons" for obligation to self-restraint and self-regulation, Baier maintains, are superior to other motives because such restraint is the sole means of avoiding the misery of unmitigated discord in an Hobbesian "state of nature." The rules to be observed for this purpose, according to Baier, are those inculcated by society and enforced by its established authority. The moral code, accordingly, will be relative to the social group.

The extreme form of this doctrine is expressed by such writers as R. M. Hare, A. J. Ayer, and C. L. Stevenson, who make ethical and moral statements purely subjective, as expressions of emotion, of feelings of approval or disapproval, or as prescriptions to which truth and falsity are irrelevant. If any criterion of goodness is recognized at all, it is pleasure, a merely subjective feeling, which varies with the individual (*de gustibus non disputandum*) so, this kind of emotive ethics cannot but be relativistic.

The same outcome follows from the unyielding separation of fact from value. G. E. Moore, in *Principia Ethica*, distinguished sharply between natural and nonnatural qualities, castigating as fallacious the attempt to rest judgments of value on the former. Bertrand Russell emphatically maintained the severance between facts and values, declaring,

> A judgement of fact is capable of a property called "truth," which it has or does not have quite independently of what any one may think about it. . . . But . . . I see no property analogous to "truth" that belongs or does not belong to an ethical judgement. This, it must be admitted, puts ethics in a different category from science.[4]

The relativism of moral imperatives is implicit in this separation of fact from value, judgments of which are implicitly held to be subjective.

The same strict separation of fact from value is reasserted by D. J. O'Connor, Karl Popper, Margaret MacDonald, and Kai Nielsen in their attacks on the traditional doctrine of natural law, all of them accusing it of committing the so-called "naturalistic fallacy"—a direct corollary of the Newtonian conception of objectivity.

The consequence of this emotivism and noncognitivism in ethics is the impossibility of establishing any rational ground for conventional morality. Moral judgment, as the expression of feeling, is to be explained by psychology and sociology. But these sciences have been assimilated to the Newtonian paradigm; their objects are treated as external facts and their pronouncements are value free. They have declared moral beliefs to be relative both to individuals, as socially conditioned, and to specific cultures, in both cases devoid of any objective criterion of "goodness" or "right and wrong." Once this thesis is accepted any rational justification for moral obligation is dissolved away; standards, being purely relative, no longer have any claim to universal respect, and the essential concept of morality is undermined. Consequently, not simply relativism, but general ethical skepticism has come to prevail.

This position in theory has been and is currently reflected in practice. Apart from such professions as religious fundamentalism, no objective standard of conduct is recognized. Permissiveness has become pervasive, and the rising generation is bewildered and deprived of direction in the conduct of life. The results are sexual promiscuity (bringing in its wake an epidemic of AIDS), organized pedophilia and other sexual abuses, the drug culture, and an alarming increase in indiscriminate violence and all manner of crime. In general, no clear idea is entertained of the end of human endeavor, which is characterized merely as "success"—although it is not specified in what respect one should seek to succeed. The current use of moral terms and apparent deference to rational precepts are generally recognized as a front for propaganda in favor of personal preferences and arbitrary partisan social ends, not only by theorists but by the public at large. Respect for persons has given way to manipulation of individuals. Professedly "ethical" political policies are vitiated by hypocritical inconsistencies, for the justification of which casuistry flourishes. For example, the U.S. administration expresses concern about the violation of human rights in China while it turns a blind eye to similar violation in Burma and Malaysia and has in the past tolerated and even abetted oppressive dictatorships in Peru and Argentina. The United Nations takes military action against Iraq for its defiance of Security Council resolutions, but it ignores similar defiance by Israel. NATO wages war against the Serbs on behalf of the Albanian Kosovars but ignores the action of the Turks against the Kurds.

In part as a result of all this, and in part its cause, is the prevalent ignorance and uncertainty about moral criteria and objectives. Not only is the rising generation bewildered about the purposes of life and prone to seek refuge in psychedelic placebos, but we are all faced with momentous and searching questions that demand immediate answers. Scientists have progressed to a point where they can perform what at one time would have been considered miracles: They can transplant organs into living beings, transform inherited characters, clone animals, and prolong life. The question in many minds whether any of these feats are legitimate requires for an answer the recognition of some moral criterion. Contemporary science has enabled nations to manufacture weapons of mass destruction. Is their doing so justifiable? Is it right to take action, even violent, to prevent their doing so (as for instance in the case of Iraq)? What should be the aims of an "ethical" foreign policy? Such questions of vital importance can only be answered if some objective standard is accepted. But today all objective standards have been abandoned, and have been denied by the scientific researches of anthropologists. When we are most in need of them, our belief in them has been eroded away.

RELIGION

The influence upon religion of the Newtonian paradigm had, as we have seen, produced a contrast and conflict between faith and understanding, engendering in the Enlightenment a wave of agnosticism. This has spread markedly during succeeding centuries, and, in the twentieth, atheism has been openly advocated by philosophers such as Bertrand Russell, Alfred Ayer, Anthony Flew, and Jean-Paul Sartre, as well as many others who have been influenced by Nietzsche, Marx, and Freud, but who have been less forthright in their rejection of religion.

Neo-Darwinians have plied the argument that all order and design in living beings can be adequately explained simply in terms of random mutations and natural selection. Any teleological argument for God's existence is thus ruled out of court. They treat mutations as occurring by chance to separable genes; so the tacit presupposition of the doctrine is genetic atomism. Daniel Dennett, for instance, who advances this line of thought most strongly in his book *Darwin's Dangerous Idea: Evolution and the Meanings of Life*, conceives chance as acting upon an imagined "Library of Mendel," which consists of a haphazard collection of nucleotides that may be shuffled at random and from which those having survival value are "selected." Selection, however, is said to be "blind"; it cannot exercise any genuine choice or preference. Once again, the atomistic and mechanistic habit of thought engendered by the Newtonian paradigm is operating here, if somewhat surreptitiously. Both in this book and in his *Consciousness Explained*, Dennett is explicitly materialistic, equating the mind with the brain with the implication of pure epiphenomenalism. His rejection of God as what he calls a "sky-hook" is, once again, a return to Laplace's Newtonianism: "Je n'ai pas besoign de cette hypothèse."

Empiricists (contemporary along with others) maintain that "existence" is not a predicate and that it is always a contingent matter whether anything exists. It follows therefore, they argue, that the existence of God cannot in the nature of the case be necessary, and the Ontological Argument for God's existence (and consequently all the other traditional arguments which, as Kant demonstrated, depend upon it) must be invalid. John Findlay on such grounds went as far as to propound an Ontological Argument for the nonexistence of God.[5]

Those who have followed the pervasive fashion of linguistic analysis have reduced religion to "God-talk"—a peculiar linguistic convention, without apparent implication of any actual reality. Norman Malcolm contrived even to defend Anselm's Ontological Argument for God's existence on the ground that contemporary logic defines logical necessity in terms of the definition of words, and Anselm had adopted a linguistic convention which entailed that the definition of God made His existence logically necessary. Malcolm's Catholic predilection apparently led him to disregard the fact that the choice of a linguistic convention is arbitrary and that nothing prevents us from adopting an atheistic convention in preference to Anselm's.

On the other hand, thinkers following Kierkegaard have turned their backs on science (as currently conceived) and have made "the leap of faith," accepting revelation without rational justfication. The opposition of faith and reason, we have observed, is a direct consequence of the Newtonian paradigm. Faith, however, is subjective and varies with the person and the sect, reinforcing the case of the relativist.

Meanwhile membership of the established denominations has markedly declined, new and fantastic cults have grown up, some of them claiming the authority of science (e.g., Christian Science and Scientology) several exerting various forms of reprehensible psychological pressure and coercion on their adherents. Religious fundamentalism has thrown reason to the winds and has produced conflict, not only between fundamentalists and liberals, but between rival fundamentalisms in different religions, amounting in extreme cases to terrorism and internecine strife.

The corrosive effect of the Newtonian paradigm upon religion, far from abating, continues. There are some who contend that only empirical science is capable of discovering truth and that every other claimant, whether religion, astrology, or philosophy, is mere superstition. Their affirmations are "all in the mind" and have no foundation in fact. Accordingly, with the decline of religious belief, the controversy among religious sects, and the prevalence of ethical skepticism, modern civilization is plunged into uncertainty, and conduct, personal as well as social, lacks direction or aim.

ECONOMICS

In economics, the prevailing ethical utilitarianism is reflected in a capitalism that has become worldwide and a market economy that prevails universally. Even where communism still persists it is giving place to more or less individual enterprise, the driving force of which is private profit. Gain and commercial advantage is the primary consideration, even when the contemplated commerce is in genetically engineered products, the dangers of which to the planetary ecology are still unknown; patents are sought for commodities that could harm humans and the environment in as yet untested ways, and there is no agreement as to the method or extent of regulation. The pre-

vailing motive is gain, and self-interest and moral considerations are ignored, or if given lip service, are argued away.

The result of this free market economy has been the growth
of transnational corporations and the accumulation of wealth
in the hands of a select minority at the expense of the majority.
The gap between the rich and the poor has widened even in
the "developed" economies, while that between the "developed" and the "developing" has brought the latter to the verge
of catastrophe. The poorest countries are paying more in interest on their international debts than their economies can sustain, while corruption in high places undermines any assurance
that even the cancellation of debts will produce the desired remedy. Concurrently the world economy wrestles with the problems of inflation, recession, "boom and bust," unstable rates of
exchange between national currencies, and the general failure
of any kind of national or international regulation genuinely
to protect the prosperity or welfare of the consumer.

Marxist communism, which sought to stem the tide of capitalist anarchy in the interest of the workers, was theoretically
tainted by the materialism inherent in Newtonian thinking and
a consequent failure of respect for individual personality and
human rights. The outcome was a totalitarianism which dis-
torted and corrupted the underlying humanistic motivation of
Marxism in its professed aim of abolishing the exploitation of
labor and brought about the eventual downfall of the Soviet
system, creating an intolerable degree of chaos and anarchy,
poverty, and mafia crime in a Russia attempting to readjust itself to a free market economy. A similar process seems to be
occurring more slowly in China. Capitalism and the market
economy increasingly prevail.

The criterion of economic success is growth, and growth is
measured by profit rather than welfare; prosperity is judged
by the standard of living of the "haves" in disregard of the plight
of the "have nots." Moreover, the assessment of growth is made
in disregard of the effects that the means of production have
upon the environment, which are more often than not deleterious. Natural capital in the form of soil fertility, water supply,
woodland, mineral deposits, and the like, is not included in

the estimate, which takes account only of manufactured capital such as machinery, plant, and financial capacity. The growth pursued is therefore unsustainable, with seriously detrimental consequences.

The spectacular success of Newtonian science led to the mechanization of production in the nineteenth century and the Industrial Revolution. This has now progressed to such an extent that vital resources are nearing exhaustion. Increase in world population and technically "improved" farming methods have caused overcultivation of land, soil erosion, and fertility degradation. The use of fossil fuels is polluting the atmosphere with poisonous and greenhouse gases. Tropical forests are being cleared for pasturage and agriculture, which, due to the resulting unsuitable soil conditions, tend to fail, leaving the land barren. The destruction of the rain forests (for this purpose and for the hardwood trade), is drastically reducing the major source of oxygen in and the absorption of carbon dioxide from the earth's atmosphere, along with a grievous loss of living species vital to the food chain and valuable for medicinal purposes. The results include spreading desertification and frequent protracted droughts in areas where subsistence conditions of life make provision precarious at the best of times. Accordingly, mass starvation recurs periodically, often exacerbated by civil conflict, which has its origin in this and other influences of the surviving Newtonian paradigm. In general the planetary environment, delicately accommodated to the support of life on earth, is being disrupted at a rate which, if not speedily checked, will (if it has not already) become irreversible.

The attitude to the natural world encouraged by Renaissance science was to regard it as a merely mechanical instrument to be exploited for the benefit of human beings. Accordingly, no sense of obligation to conserve the environment was fostered either in the prevailing moral consciousness or by economic practice. Today the effects of this attitude have been to exhaust and destroy the very means that world capital seeks to exploit, and, by polluting the environment and disrupting the ecology, threaten the very survival of life itself upon the planet.

POLITICS

Eighteenth-century ideas still dominate twentieth-century politics. The concept of individual rights may have a sound foundation, but the dependence of the individual upon society is rarely recognized as it should be, so that individualism is dominant in such political theory as has survived the attacks of positivism and has escaped the stigma of ideology. This asocial, individualistic, conception of rights predominates in the thought of such writers as Nozick and Rawls, both of whom (even where they disagree) maintain a form of contract theory as the original source of social justice.

In practical politics, the accepted standard of good government is that of liberal democracy, in which government intervention is reduced to a minimum, and individual freedom is extolled. The conduct most admired is that of the entrepreneur who seeks personal success—each individual for himself and the Devil take the hindmost. The presumed function of government is to hold the ring, and the provision of social welfare has to be limited in order to keep public expenditure in check and to reduce taxation. Political parties aim rather at gaining power than at public service, and so their leaders pander to popular demands for low taxation and freedom from interference, in disregard of genuine social justice. What the public wants takes precedence over what society needs, and the pursuit of individual advantage overshadows the common good. The consequence is popularism and official corruption—mostly surreptitious, but no less harmful. At the same time, in order to court popular approval, politicians prevaricate in their public statements and give evasive answers to pertinent questions, so that people, especially the young, increasingly mistrust them and tend to stay aloof from politics, neglecting to vote at elections. This dilutes the democratic character of the administration and the rift between government and the governed is continually widening.

In consequence, insistence on individual freedom has come full circle and is undermining the very principles of genuine democracy. Weighty issues affecting the common welfare are decided in elections by popular vote, that is, by a majority for the most part ignorant, self-interested, and least qualified to judge, while the power of economic vested interests is enhanced by the ability of lobbyists to give financial support to political parties.

While democracies suffer from these prevailing ills, dictatorships violate human rights and embark upon exaggerated economic schemes that are fraught with a prospect of environmental disaster and severe social disruption (for example, the building of dams to flood vast areas for hydroelectric generation in China). In developing countries, the ruling powers, taking their cue from the more developed, meet any attempt at liberalization of government with opposition, and upset attempts to establish democratic institutions by military coups to set up military dictatorships, which impoverish the people and enrich the dictators (Indonesia and Nigeria are but two examples).

The most dangerous effects of atomistic and separatist thinking are to be seen in the theory and practice of international affairs. Here the dominating concept is national sovereign independence, a status demanded by every ethnic group and cherished by every national state. The claim of independent national states to sovereignty creates an impasse in world affairs that is seldom diagnosed and presents so serious a problem that discussion of it merits a chapter to itself. Like everything that has been detailed earlier, this persistent way of thinking that independent sovereign states are the only possible and natural political structure for the nations of the world is a hangover in the twentieth century of the world view typical of the seventeenth and succeeding centuries.

It is now imperative to shake off this outdated mental bias that is leading world civilization to self-destruction. The new, twentieth-century, scientific paradigm has replaced that of classical Newtonian physics and, could its implications be realized and take hold in other spheres than physical science, the paradigm would offer a fresh approach to world problems in ways that we shall consider at more length, after we have examined in some detail the formidable obstacle presented by national sovereignty.

NOTES

1. Cf. E. E. Harris, *Formal, Transcendental and Dialectical Thinking* (Albany: State University of New York Press, 1987), chapter 2.

1. Cf. Bertrand Russell, *Human Knowledge* (London: Allen and Unwin, 1948), pp. 374, 490; idem, *The Problems of Philosophy* (London: Oxford University Press, 1950), chapter 6.

3. Cf. Kurt Baier, *The Moral Point of View* (Ithaca, N.Y.: Cornell University Press, 1958).

4. Bertrand Russell, "Reply to Criticisms," in *The Philosophy of Bertrand Russell*, ed. P. A. Schilpp (Evanston, Ill.: Open Court, 1946), p. 723.

5. Cf. J. N. Findlay, "Can God's Existence Be Disproved?" in *New Essays in Philosophical Theology*, ed. A. Flew and A. MacIntyre (New York: Macmillan, 1955), pp. 54–55.

4

The Problem of Sovereignty in International Relations

In the Western world, theories of sovereignty were set out by political theorists at roughly the same time as the rise of the national state in the fifteenth and sixteenth centuries. This was also when Copernicus published his revolutionary suggestion that the sun was the center of the solar system and not the earth. The theoretical concepts and devices they used were Natural Law and original contract, ideas inherited from the Middle Ages and derived from Roman law, but now modified in accordance with the paradigm of Renaissance science. The assumption was generally made that, prior to the civil condition, people lived in a state of nature and emerged from it by contracting to give up their natural rights and to live under a common law. But these concepts were not used by all thinkers to produce the same conclusions.

There were two main types of theory, one (which was espoused by the royalists in the English Civil War) was inspired by the Catholic tradition of Divine Right and embraced by the inheritors of the Holy Roman Empire; the other was inspired by the Reformation and advocated by those rebelling against despotic rule in The Netherlands and in England. The first, typified by the writings of Jean Bodin and Thomas Hobbes, stressed the supremacy and absolute power of the sovereign; the second, exemplified by the works of Althusius, John Milton, and John Locke, insisted on the rights of the subject and the necessity for the consent of the governed. Each type of theory drew attention to an essential feature of government in modern civilized society: the first, to the indispensability of a final court of appeal with unchallengeable power to enforce the law; the second, to the requirement that the exercise of sovereign power be legitimized by its service of the common interest and authorized by the consent of the governed, if the sovereign body is to retain its ability to govern. These two aspects of sovereignty I have elsewhere referred to as the juristic and the ethical, respectively.[1] They also reflect characteristics of the new scientific paradigm introduced by Copernicus: the isolationism consequent upon atomistic thinking, which kept sovereign powers separate and mutually independent; and the individualism that regarded persons as each claiming by natural right an inalienable circle of rights to be defended against encroachment from the demands of others and especially of the ruling power.

For Hobbes human beings were by nature self-seeking, acquisitive, and aggressive. In the state of nature therefore they were constantly at enmity and in conflict. In this condition of fear and insecurity, life would be intolerably hazardous, and there would be no assurance of possession or any agreeable amenities. Individuals therefore would be led by reason to contract with one another to give up their natural powers to a single person (or group of persons) who would rule over them as sovereign with supreme power.

The legalistic notion of contract presupposes that the contracting parties are independent persons, each pursuing interests peculiar to him- or herself, reflecting again the atomistic approach encouraged by the Renaissance paradigm. We have noticed the same tendency in economic thinking as it emerged in the work of Adam Smith.

Theorists who insist upon the juristic aspect of sovereignty all agree that states, being subject (under God) to no higher authority, are in their mutual relations in the state of nature; hence they are potentially hostile one to another and perpetually in a posture of antagonism and war, which, as Clausewitz put it in the nineteenth century, is simply politics conducted by other means. Hobbes maintained that

In all times, Kings and Persons of Soveraigne authority, because of their Independency, are in continual jealousies, and in the state and posture of Gladiators; having their weapons pointing, and their eyes fixed on one another; that is, their Forts, Garrisons, and Guns upon the Frontiers of their Kingdoms; and continual Spyes upon their neighbors; which is a posture of War.[2]

Spinoza, likewise contended that

Two states are enemies by nature. For men in the state of nature are enemies; and so all who retain the right of nature outside the state remain enemies.[3]

And Hegel was of the same opinion:

Since the sovereignty of a state is the principle of its relation to others, states are to that extent in a state of nature in relation to each other. . . . It follows that if states disagree and their particular wills cannot be harmonized, the matter can only be settled by war.[4]

Thinkers of the other school, who give precedence to the ethical aspect of sovereignty, limited the internal power of the state by what they saw as the Law of Nature, and sought, like Grotius, to recognize or devise a means of regulating the behavior of states one to another, according to a natural law above and outside of their own authority. This is the origin of what today we call International Law.

Unquestionably, any form of social organization implies the regulation of the conduct of its members by laws tacitly recognized or publicly promulgated, customarily observed or administratively enforced. In every such society there will be those who, in their perception of their own interests, seek to break such rules, so that some method of coercion is essential to the good order of society. Moreover, laws are liable to different interpretations, and resort to an impartial authority is needed whenever disputes arise. The settlement of disputes and the prevention of antisocial conduct require a supreme authority to whom final appeal can be made and who has power, unchallengeable by private individuals, to enforce the law. This is the legal sovereign, and this is the rationale of the juristic theory. Sovereignty so conceived is indivisible, because if it were shared by different bodies, there would have to be a superior body to decide between them when their actions or decisions were in conflict—and that would be the ultimate sovereign. It is also inalienable, because in whomsoever it is vested is the supreme authority above which there is none capable of transferring it to another. Nor can the sovereign limit or renounce its own status, which will remain absolute even if power is delegated to nominated officials.

On the other hand, power to govern a community always exceeds the physical capability of mere individuals, and those who wield it can do so only with the cooperation and consent of a large body of the people over whom they exercise it. What authorizes their rule over the social group must be their ability to maintain the welfare and retain the confidence of their subjects. Their power qua political power is always derived from the community over which it is exercised and is justified by the protection it affords to the rights of its subjects as citizens and the extent that it ministers to their welfare as persons. Even rulers who rely solely on military force can do so only as long as their soldiers remain loyal to them (a loyalty that will not last if men are treated merely as cannon fodder), and as long as their armaments and supplies can be provided by the industry of their subjects (on which they can rely only as long as the majority are content with their lot). The ethical theory of sovereignty, consequently, also has a strong measure of truth. Both these theories, as they were advocated in the sixteenth and seventeenth centuries, were based upon an individualistic conception of subject people and an isolationist view of sovereign states, concepts which still underly political thinking even today.

In fact, both of these aspects of sovereignty are combined in practice, and some theorists reconcile them in a conception which sees the will of the people as absolute. Spinoza writes,

Where men hold rights in common, and are all led as if by one mind, certainly . . . each of them has the less right the more the rest are together more powerful than he, that is he has no actual right in nature except what the common law (*jus*) concedes to him; for the rest, whatever he is ordered to do by common consent he must carry out, or is compelled to do by right. This right, which is defined by the power of the multitude, is called sovereignty (*imperium*). And he who by common consent has the care of the republic holds this [power] absolutely.[5]

Similarly, Rousseau, who with Hobbes recognizes that sovereignty is absolute, inalienable, and indivisible, concludes that it belongs only to the people as a whole, or what he calls the General Will. It issues from what he conceives as the Social Contract, by which

Each of us puts his person and all his power in common under the supreme direction of the general will, and, in our corporate capacity, we receive each member as an indivisible part of the whole. . . .

This public person, which is formed as if by the union of all the others . . . is called by its members *State* when it is passive, *Sovereign* when it is active.[6]

Nevertheless, the notion of individuals as separate, each holding a closed circle of rights, persists, along with that of states as independent and supreme each within its own boundaries.

These theories, resting on tacit presuppositions typical of the sixteenth and seventeenth centuries, have been restated, criticized, and reformulated by more recent philosophers, but in principle and in essentials they have not been superseded. Not much has been written about sovereignty since the 1920s and 1930s that has done little more than reaffirm one or the other of the main theories. If the conduct of states up to the present time is observed, it is found to bear out very closely the principles set out by the major thinkers. Governments claim an absolute right to legislate and administer their own laws, and even where individual liberties are demanded and conceded, they can be made effective only as recognized and protected by the law enforced by sovereign institutions (executive and judicial). Yet those who govern can retain power only as long as the people over whom they rule acquiesce and in sufficient measure submit. If en masse the population resists the party in government, that party will forfeit its power. In democratic countries, this happens at election time; in totalitarian regimes, by bloodless or violent revolution. This has been borne out by events in the Soviet Union and her satellite countries in Eastern Europe, in South Africa, and in Latin America.

In international affairs the relations between states up to the present time, despite the appearance of organizations such as the League of Nations and the United Nations, have not really changed in principle from what they were said to be by Hobbes, Hegel, and Clausewitz, and as long as the several nations continue to claim and to be recognized as sovereign, this cannot be otherwise. Because sovereignty is indivisible, the proposition often made that today the sovereignty of nations is limited by their obligations under International Law and to the United Nations is unacceptable, for reasons presently to be noted. Observance of International Law and deference to the United Nations are always conditional on the policies and sovereign decisions of the governments in power. In consequence, national sovereign independence is a persistent obstacle to the maintenance of world peace and the conservation of the global environment. The problem which it presents is rarely recognized, yet unless it is faced, the prospect for mankind in the twenty-first century is likely to be extremely bleak.

It is the juristic aspect of sovereignty that is chiefly in evidence in international relations. The nations each claim independence, the recognition of which they demand from others. They protest in the strongest terms against outside interference in what they claim to be their internal affairs. They seek to defend themselves at all costs against such interference of whatever kind. Over such internal affairs they claim absolute jurisdiction. And they pursue exclusively and assiduously what they regard as their national vital interests.

The common pronouncement of politicians daily confirms this assessment. One has only to read (or hear on the radio and television) the speeches of British politicians in connection with European integration, particularly the more Euroskeptical. British national interests, equated with the maintenance of British sovereignty, are always uppermost in their minds. The same is true of most politicians in other countries of Europe, even those who are in favor of closer unification. The statements of the American president and of members of Congress leave one in no doubt of their single-minded defense of American interests, be they in defense or in economic prosperity. The developing nations are equally insistent on being recognized as sovereignly independent; and the troubles in the former Yugoslavia, in the Commonwealth of Independent States, the erstwhile Soviet Union, and the less menacing tensions in the former Czechoslovakia, are all the result of the claims to sovereign independence of the several ethnic groups. A similar claim by Albanians in Kosovo has precipitated the worst crisis in Europe since World War II.

The ethical aspect of sovereignty is implicit in the priority given by governments to national interests, for what is seen as of vital interest is supposed to be whatever serves the common welfare of the nation, the efficient protection of which is what entitles the government in office to exercise sovereign power. National interest is simply individual interest transferred to the political group to which the individual belongs. Whether nowadays the true common interests of the people in the several nations coincide with what their present governments hold to be their national interests is a matter further to be considered; but in practice governments invariably behave as if they do.

The Statute of the International Court of Justice (repeating a similar clause in that of the former Permanent Court of International Justice) lays down that only sovereign states can be subjects of International Law. The definition of sovereignty in International Law, however, leaves one in no doubt as to the absolute supremacy and independent status of its presumed subjects.

Dr. H. Lauterpacht has expressed the matter thus: "The sovereign State does not acknowledge a central executive authority above itself; it does not recognize a legislator above itself; it owes no obedience to a judge above itself." In the judgment on the Palmas case in 1928, Judge Huber made the pronouncement that

Sovereignty in the relation between States signifies independence. Independence in regard to a portion of the globe is the right to exercise therein, to the exclusion of any other State, the functions of a State. The development of the national organization of states during the last few centuries and, as a corollary, the development of International Law, have established this principle of the exclusive competence of the State in regard to its own territory in such a way as to make it the point of departure in settling most questions that concern International Relations.[7]

Advising the League of Nations in the Eastern Karelia case in 1923, the Permanent Court of International Justice stated that the recognition of sovereign independence of states was "a fundamental principle of international Law."

But if sovereignty is defined in law as it has been by Dr. Lauterpacht, the authority of International Law is immediately cancelled out, for the sovereign state cannot remain sovereign and acknowledge a superior legislation to its own and cannot therefore be subject to International Law. Similarly, if the state owes no obedience to a judge above itself, it owes none to the International Court of Justice. In actual practice, that court can only deliberate on cases where the parties involved have consented to accept its decision, and its judgment may be rejected by any of them if it is not to their liking.

Consequently, the authority of International Law is fictional, even in theory, and its precarious nature is even more apparent when one notices that it is the product of no established legislature, but rests only upon treaties, such as the Hague Conventions (usually referred to as law making treaties). Now, treaties are said to be sancrosanct, and international lawyers lay down the principle, *pacta sunt servanda* (treaties must be kept); yet in practice they are kept only when they are believed by the states that have entered into them to be in their national interests. When national (or so-called "vital") interests are not served by observance of their terms, treaties are invariably renounced or ignored. In any case, states, being sovereign, are at liberty to interpret a treaty in whatever way best suits their interests; they are also free to retract their commitments whenever they deem the circumstances warrant. Presidents Theodore Roosevelt and Woodrow Wilson both maintained that a nation could renounce a treaty at any time it thought fit; and W. E. Gladstone speaking in the House of Commons in 1870 denied that "the simple fact of the existence of the guarantee is binding on every party of it, irrespective altogether of the particular position in which it may find itself at the time that the occasion for acting on the guarantee arises." Numerous other statesmen have expressed similar views, and the prevailing conduct of states bears them out.

The pages of history are littered with accounts of broken treaties. A few examples may be cited. In 1668 Charles II of England concluded a treaty of alliance with Sweden and The Netherlands, but four years later he joined with Louis XIV in war against Holland (having signed the Secret Treaty of Dover with France in 1771). In 1818 the Quadruple Alliance of Russia, Prussia, Austria, and Britain solemnly declared an "unchangeable resolution never to depart, either among themselves or in their relations with other states, from the strictest observations of the Principles of the Law of Nations." But in 1831 the czar suppressed the Kingdom of Poland which had been set up at the Congress of Vienna by a treaty of which Russia was a signatory. Later Britain destroyed a Turkish fleet without any declaration of war, and Prussia overran the Duchies of Schlezwig and Holstein without pretext or title. The rest of the nineteenth century is a catalog of similar breaches of International Law and treaty obligations, which culminated early in the next century with Kaiser Wilhelm's tearing up of "the scrap of paper" which committed him in 1914 to respect Belgian neutrality. Neither the Treaty of Versailles nor the League of Nations could prevent Adolph Hitler from reoccupying the Saar or from annexing Austria; nor could the agreement he signed in Munich

with France and Britain restrain him from overruning Czechoslovakia in 1938.

The reason for this catalog of perfidy is plain. It is that the
nations are sovereign and always exercise their sovereign pre-
rogative of acting in what they perceive as their national interests. As sovereign, they are subject to no higher authority, and
there is no way of enforcing the observance of a treaty on a
sovereign government except the threat of *force majeur*. Hegel
therefore made no mistake when he argued that the fundamental principle of International Law (that treaties ought to be kept)
"does not go beyond an ought-to-be (*bleibt daher beim Sollen*)."[8]

Some might contend that since World War II and the establishment of the United Nations, all this has changed, but it is
not so. The Charter of the United Nations is itself no more than
a treaty, and it commits the organization in Article 2 to the principle of the sovereign equality of all its members. Consequently,
the resolutions of the Security Council have been ignored time
and again by South Africa, Israel, North Korea, and Iraq, to
mention only these, while the permanent members have been
able to veto any decision that did not seem to them compatible
with their national interests. If any member chooses to ignore
or defy decisions agreed in the General Assembly or in the Security Council of the United Nations, because it is a sovereign
state, it cannot be forced to comply except by some form of
military threat.

Any such coercion is ultimately military, because economic
and other seemingly nonmilitary sanctions cannot be made effective unless they are backed by military force. If they are to
"bite," they have to be universally applied, so some form of
pressure is needed to persuade members in general to impose
them. Yet sanction busting commonly occurs if no military provision is made to prevent it; frequently (as was the case with
respect to Haiti, and has been threatened in the case of Serbia)
a naval blockade has to be mounted and forcible measures are
needed to counter attempts by the country, upon which the
sanctions are directed, from using its own military might to
break them. In short, the only effective sanction is war, in one
form or another. But international order requires the mainte-

nance of peace, and its condition is the rule of law, of which military conflict is a practical breach.

It follows that the United Nations is not equipped and is not competent to maintain world peace, for it can enforce its resolutions on its members only (in the last resort, if at all) by waging war, as it has done in the Congo, in Korea, and in Kuwait and Iraq. Its attempts at pacification in cases of civil war in Bosnia, in Kavorno–Karabakh, and in Chechnya have been hopelessly frustrated by its obligation, imposed by its charter, to respect the sovereign rights of its members. In cases of dispute it cannot adjudicate (witness the futility of the efforts made to mediate between the warring parties in the former Yugoslavia). If it could, its judgments could not be enforced. Its agencies are obliged to observe strict neutrality and are dependent on the consent of the warring sovereign nations in order to function, so that their operation is constantly frustrated, as it was for instance in Bosnia. Add to this that the sovereign members of the United Nations (always giving preference to national interests) are chronically reluctant to supply sufficient funds, and the frequent failure of its agencies, despite the devoted and skilled efforts of their employees, is hardly surprising.

The ineptitude and illogicality of the international set-up has been clearly illustrated by the debacle that has occurred over Kosovo. The brutal policies of Slobodan Milosevic of ethnic cleansing were not to be tolerated and were rightly condemned by the so-called international community. The United Nations could not take effective action because of the opposition of Russia and China, whose veto in the Security Council was feared by the other members. NATO therefore tried to impose a relatively humane settlement on Serbia, which she rejected, leaving no option but the use of military force, which had been threatened as the sanction against continued intransigence, and then had to be carried out, lest "credibility" should be sacrificed. The air attacks that followed only made matters worse, as should have been expected. They consolidated support for Milosevic at home, even among his former opponents, and gave him the pretext he needed to drive out the Albanian Kosovars with increased ferocity, creating the most serious refugee crisis in Europe since World War II, one that amounted to a humanitarian disaster. The aerial bombardment, precision bombing notwithstanding, inevitably caused civilian casualties, and the destruction of the Serbian infrastructure affected not only that country, but, by cutting her off from her neighbors, ruined their commerce with Yugoslavia. The accidental bombing of the Chinese embassy in Belgrade caused a diplomatic crisis and rendered Chinese consent to any settlement of the conflict by the United Nations highly precarious. Serbia sought to have the action of NATO condemned as illegal by the Court of International Justice, but apart from the unconscionable time required to get a ruling from the court, its jurisdiction can be repudiated at the discretion of the parties concerned. The result has been an appalling situation in which every course of action taken has been wrong, and no reasonable solution can be envisaged. Even when the Albanian Kosovars were eventually allowed to return, their homes and villages were destroyed, they lost all their possessions, thousands of their menfolk had been massacred, and their suffering has been traumatic in the extreme.

The inadequacy of the United Nations is moreover no more than symptomatic of the endemic disease of international politics with which it is infected by national sovereignty. Because every nation is fully aware that there is no superior power to protect its sovereign rights and that its independence can be ensured only by its own military strength, the primary vital interest of every national state is that of security, which of necessity comes to take precedence over all others when national interests are considered. Accordingly, the first and most insistent demand on a nation's resources is defense. It seeks to maintain such military capacity as it can and to augment it by means of alliances with other nations whose national interests are compatible with its own. Those whose main interests are opposed see one another as potential enemies, with the result that rival blocs are formed and there is a persistent effort to maintain a balance of power.

But this balance is very unstable; nothing more is needed to upset it than a technical breakthrough in weapon efficiency on either side. As each bloc, suspicious of possible build-up by its opponent, is constantly seeking to strengthen its own potential, an arms race is inevitably generated, tensions build up, and crises intermittently recur, threatening or actually breaking out into armed conflict. All this is reflected in the history of diplomacy and in the proceedings of the United Nations; and the overall results are periods of so-called peace, interrupted by frequent crises and minor wars, which threaten to escalate into major warfare. The destructive capacity of modern weapons makes this pervasive threat an intolerable prospect; yet, as we have seen, the United Nations is powerless to counter or to mitigate it, and there is no other means to remedy the inevitable effects of rivalry between sovereign states whose competing national interests cannot be reconciled.

The history of the past three centuries nicely illustrates this pattern of recurring warfare. After a succession of European conflicts and colonial wars in the eighteenth century, the Napoleonic wars engulfed Europe and had worldwide repercussions. The nineteenth century, despite expressed intentions by the great powers to maintain peace, was marked by a succession of minor wars and crises, ultimately culminating in the outbreak of World War I—"the war to end all war." Alas! The Peace of Versailles might better have been described as "the peace to end all peace." The League of Nations made little if any difference to the succession of crises and minor wars during the next twenty years ("the twenty years' crisis," as E. H. Carr called it in his book of that name), which led to even greater and more devastating conflagration, World War II. Since 1945, the establishment of the United Nations notwithstanding, there have been more than 150 minor wars, some of them waged in the name of the supposedly peacekeeping organization. The Cuban missile crisis brought the world perilously close to a third world war, and the succession of crises continues to the present day without prospect of final settlement, even although the Soviet Union has collapsed and the Cold War has allegedly ended. Although the arms race between the superpowers has temporarily somewhat abated, it persists among the developing countries, in particular India and Pakistan, Israel and her neighbors, especially Syria and Iraq.

The oft-repeated opinion that the policy of so-called nuclear deterrence has prevented a major war for the past fifty years is decidedly misled and misleading. Nuclear deterrence is no more than the contemporary form in which the persistent attempt to maintain the balance of power is being pursued. It has now become a balance of terror, the strategy appropriately acronymized as MAD (Mutually Assured Destruction), the instability of which until recently was rather augmented than decreased by the continued build-up of nuclear arsenals. American strategists at one stage were entertaining the idea of the possibility of winning a nuclear war and may even have been projecting plans for a preemptive first strike. With the end of the Cold War, the superpowers have agreed to destroy their more obsolete nuclear weapons, but this alone has created new problems, not only the difficulty of disposing of nuclear waste, but also the possibility that the states of the former Soviet Union, to acquire much-needed hard currency, might sell materials and know-how to smaller nations aspiring to nuclear capability; or that organized crime might succeed in supplying the means of manufacturing atomic bombs to terrorists. The major powers retain their nuclear arsenals, and the United Nations can do little or nothing to stop nuclear proliferation. The menace of nuclear war has not yet been averted. Almost a hundred years elapsed after the Battle of Waterloo before Europe was plunged into another Great War. Since the end of World War II, we have enjoyed little more than fifty years of unstable and somewhat spurious "peace," and the possibility of another major conflict is still with us.

As the twentieth century draws to a close, over and above the difficulty of maintaining peace, mankind is facing enormous problems arising from the destruction of the environment. The inordinate growth of population and the consequent demand for food and industrial products the world over have generated widespread interdependent and mutually exacerbating problems. The earth's resources are being used up at an increasing rate. Consumption of fossil fuels in vast quantities is polluting the atmosphere and creating a greenhouse effect which threatens climatic change of dramatic and catastrophic proportions. The resulting loss of food crops is likely to decimate the population in many areas. The widespread destruction of tropical rain forests, the main source of atmospheric oxygen, has removed a major means of reducing atmospheric carbon dioxide, which is the chief greenhouse gas contributing to global warming. Depletion of the forests is also a contributory cause of desertification, the effects of which are cumulative, putting the survival of wildlife, livestock, and people at risk. Further, the use of CFCs is depleting the ozone layer that protects living things from excessive ultraviolet radiation, exposure to which causes cancers in humans and is fatal to some food crops. Add to all this the accumulation of toxic and hazardous waste from industrial and nuclear-power production, the disposal of which (some of it remaining lethal for thousands of years) is posing a problem for which there is as yet no solution, and it is evident that the immediate future of mankind is fraught with the utmost danger.

  Associated with many of these conditions is the current atrocious loss of species, both animal and vegetable. Overfishing has depleted many fishing grounds of their stocks to an alarming extent. Along with the ravaging of the rain forests, the destruction of wetlands and the loss of wilderness to "development" is depriving innumerable species of their natural habitat. As a result, numerous species of bird life, reptiles, and mammals, as well as plant life and insects, are threatened with extinction. Hunting and poaching are so widespread that the extermination of tigers, elephants, rhinoceros, and other mammals is imminent, while the damming of rivers (for hydroelectric power) is isolating salmon and other species of fish from their natural breeding places. The pollution of the oceans is endangering the plankton, the basis of the whole food chain. Thus the entire terrestrial ecology is being disrupted; and as the biosphere is a single biocoenosis, every species of life is under threat.

  If this erosion of the environment, of which I have given only the barest outline, is not checked and reversed, the prospects for survival of life on earth will become increasingly slim, and the remnant of the human race, if any does manage to survive, will be reduced to a desperately low standard of living.

Unless effective measures are taken to counteract these threats to the environment, scientists have told us, it may well be too late by the end of the century to reverse the deleterious trends. Yet international agreement, even what could be reached in 1992 at the Rio Summit, has achieved nothing adequate. Nonbinding decisions indicate no more than failure of resolve, and the decision to limit CO_2 emissions to present levels by the year 2005 is pathetically insufficient. The reason for such reluctance to act, the so-called lack of political will, is clearly that the nations give precedence to their national interests over environmental requirements. Agreements either cannot be reached at all or, when they are made, are nonobligatory, at best half-hearted, and in any case unenforceable.

The problems are global in scope, and only global remedies will suffice to solve them. They cannot be met by private efforts or the exertions of nongovernmental organizations, because they must be legislated and be compulsory for everyone. The jurisdiction of national governments is restricted to their geographical frontiers and is thus too limited; for, even if they decide to legislate, the failure of one nation to act (e.g., with respect to air pollution or to the destruction of rain forests) is enough to nullify the measures enacted by others. Only concerted international action would be adequate to produce the necessary result. But no diplomacy can bring about concerted international action, for the reasons already stated, that sovereign nations give precedence to their own national interests, so that such agreement as can be reached is never sufficiently stringent or far reaching, and observance cannot be enforced, except by military means such as defeat the object of the exercise by increasing the damage to the environment. Nor can the United Nations do anything to ameliorate this situation, because it is committed by its charter to perpetuate it.

National sovereign independence therefore has proved fatally inimical to the solution of world problems. Yet it is on the resolution of these global difficulties that the welfare of peoples and the very survival of humankind depends. The national sovereign state can no longer effectively protect its citizens from devastation in war, nor can it protect their living standards and maintain the amenities of life in the face of environmental deterioration. The insistence of national governments on their sovereign rights cripples the ability of the United Nations to ensure either peace or environmental conservation. In short, the national state now lacks the one and only justification for the exercise of sovereign power, the fostering of national prosperity and security. Its ethical character has been undermined, and its title to be juristically supreme is no longer valid.

This fact is seldom recognized or acknowledged either by politicians or by the general public, who are still impelled subconsciously by latent Newtonian presuppositions to think in terms of isolated and independent national units. Scientists and others who see the problems and recommend measures to counteract them fail to realize that such measures require political action which, if taken only by national governments, will inevitably be insufficient in scope, and which will not be taken by national governments because they do not regard them as serving vital national interests. For the same reason, the concerted international cooperation required will not be forthcoming, and even if it were, it could not be relied on for lack of the means to enforce treaty obligations. The scientists and nongovernmental organizations like Greenpeace and Friends of the Earth may protest and demonstrate, but they cannot legislate. They can draw attention to the problems but they can do nothing effective to remedy the evils. Unless and until this problem of sovereignty is squarely faced and addressed, the prospect for civilization will be bleak in the extreme.

As long ago as 1939, Professor George Keeton wrote with justice that "the fetish of national sovereignty assumes the shape of the evil genius in the . . . forest of international intercourse."[9] The national sovereign state has become obsolete in the circumstances of the present-day world. The philosophical basis of its right to supreme power has been eroded away, and its legitimacy has been undermined by its loss of competence to protect its citizens either from military destruction or from environmental calamity.

If the decline and fall of civilization, not just in the Western world but the whole world over, is to be averted, some new

form of world government is essential, a form of which the
United Nations falls short because its professed objectives are
constantly obstructed by the sovereign rights claimed by its
members, rights which its charter endorses and protects. This
fundamental contradiction must somehow be removed. But
current efforts at reform, such as recommending increase of the
number of permanent members in the Security Council or
popular election of a new body to advise the General Assembly, are futile as long as the provision remains in the charter
that the sovereign independence of members is to be upheld.
For the new assembly contemplated would have a merely advisory function, and that only to the General Assembly, which
is itself only an advisory body. Neither have legislative power,
nor can any assembly or council, as long as sovereignty is retained by national governments. Even the abolition of the right
of veto in the Security Council would be unavailing, as long as
states are able to defy the council's resolutions with impunity
or can be forced to respect them only in the last resort by military means. If any path to salvation is to be found, something
more radical and more far reaching must be envisaged than
these so-called reforms, because they ignore the primary problem to be tackled: that of national sovereignty.

We must somehow nullify and escape from the parlous state of
affairs resulting from persistent thinking typical of the seventeenth-
century scientific paradigm. The mechanical and atomistic conception of nature that underlay Hobbes's political analysis and
the individualism that inspired both his doctrine and Locke's
notion of natural rights are the original, yet latent, sources of
modern claims, on the one hand to individual and ethnic liberty and on the other to national independence, fomenting various forms of terror and mayhem, civil conflict, and warfare, in
an era when no individual freedom is unconditioned by social
provision and regulation, no nation is actually independent of
others (either economically or culturally), and nature is no mere
external machine to be exploited at will, but the very substance
and sustenance of human economic and social existence, to be
nurtured and protected at all cost.

Historically it is to be expected that a new scientific paradigm will be very slow to penetrate lay thinking and to take effect in social practices other than in science itself. That twentieth-century physics has so far had little influence on philosophy, morals, politics, and economics is therefore to be anticipated. But the transformation of world civilization by the developed technology of modern scientific culture is now destroying the very conditions of survival for life on earth at such a rapid pace that we cannot safely wait for the normal historical processes to take their course. It is in this sorry state that humanity finds itself at the present time, under the persistent influence of Newtonian concepts, and unless human habits and ways of thinking are radically modified in short order, our entire race is in dire danger of extinction. Human conduct and way of life has in the past exterminated several interesting and beautiful species of birds and mammals and is now endangering many others—most significantly, our own.

NOTES

1. Cf. E. E. Harris, *The Survival of Political Man* (Johannesburg: Witwatersrand University Press, 1950), p. 35f; idem, *Annihilation and Utopia* (London: Allen and Unwin, 1966), p. 37; idem, *One World or None* (Atlantic Highlands, N.J.: Humanities Press, 1993), p. 44.

2. Thomas Hobbes, *Leviathan* (Oxford: Clarendon Press, 1943), chapter 3.

3. B. Spinoza, *Tractatus Politicus*, chapter 3, p. 13.

4. G.W.F Hegel, *Philosophie des Rechts*, sections 333–334, trans. T. M. Knox, *Hegel's Philosophy of Right* (Oxford: Clarendon Press, 1953), pp. 213–214.

5. Spinoza, *Tractatus Politicus*, chapter 2, pp. 16–17.

6. J. J. Rousseau, *Du contrat social* (Leipzig: Gerhard Fleischer, 1818), chapter 6.

7. H. Lauterpacht, *The Function of Law in the International Community* (Oxford: Clarendon Press, 1933), p. 64.

8. Hegel, *Philosophie des Rechts*, section 333.

9. George Keeton, *National Sovereignty and International Order* (London: Stevens, Peace Book, 1939).

5

The Twentieth-Century
Revolution

Since Max Planck's discovery of the quantum of action and Albert Einstein's formulation of the special and general theories of relativity in the early years of the twentieth century, physicists' conception of the natural world has been revolutionized. A new paradigm has been adopted which, while in some respects it has proved to be, as Max Planck maintained, "the completion and culmination of the structure of classical physics,"[1] in the main it contradicts and cancels out key concepts of Renaissance science.

What Planck meant by this quoted remark was that the new concepts removed from the classical physics certain contradictions that had arisen in the attempts to apply classical theories to newly observed phenomena (for example, the advance of the perihelion of Mercury, which Newton's law inaccurately predicts), and, by generalizing certain classical concepts, had made physical theory more coherent. As Einstein and Infeld express it, "The new theory shows the merits as well as the limitations of the old theory and allows us to regain our old concepts from a higher level."[2] This, however, is the purpose and effect of every scientific revolution. As we have already noted, revolutions are prompted by contradictions in the prevailing conceptual scheme, and the new paradigm is accepted when it is shown to be self-consistent and universal in its application.

The most troublesome contradiction in the classical physics as it had developed up to the end of the nineteenth century was the mechanical conception of the ether. As it had to offer no resistance to the passage through it of solid bodies, the ether had to be considered extremely fluid, but to be the medium of electromagnetic vibration, it needed to be exceedingly rigid. Further, the world of the classical mechanics was occupied by separable particles moving under the influence of impressed forces in absolute space, where their true places were taken to be immutable points. Position and motion, however, can be measured only with reference to some reference frame, in practice usually the earth or the sun. But both of these are in motion, the earth in its orbit around the sun, and the sun in an orbit around the center of the galaxy. Late nineteenth-century physicists were therefore at pains to discover an absolute frame in which to make measurements of electromagnetic radiation. They assumed that this would be the lumeniferous ether and that bodies (like the earth) moving through the ether would create an ether wind. The Michelson–Morley experiment was designed to measure its velocity, but the result of the experiment was null, demonstrating that there was no ether wind.

At the same time the absence of any discrepancy in the appearance of rotating double stars proved that the velocity of light is not affected by that of its source, and the Doppler and Fizeau effects showed that ether was not dragged along with a moving observer. Accordingly, the velocity of light is constant whatever that of the observer might be. This being the case, different observers using light signals to measure the distance of distant objects will find different events to be simultaneous. In short, absolute simultaneity at a distance cannot be determined. To reconcile the contradictions that considerations like these occasioned in classical physics, Einstein worked out the special theory of relativity.

He realized that measurements of length and duration will vary relative to the velocity of the reference frame in relation to the object measured. Measurements of space and time differ with the velocity of the observer, relative to the object observed, upon these again depend quantitative estimates of gravitation and mass. Relations of duration, distance, mass, and energy are all mutually involved, each being relative to all the rest. The relations are internal, determining the nature and magnitude of their terms. These inescapable interrelations are moreover inseparably dependent on the motion of the observer, who is now as important a factor in what is observed as the object of observation.

As electrodynamic theory developed in the latter half of the nineteenth century, the concept of the field took precedence over that of the particle. A charged particle or a magnet is surrounded by a field of force, that is, a configuration of lines of force, along which a free body will be accelerated. In contemporary physics, field becomes the important concept into which the particle more or less dissolves away.

Because simultaneity at a distance is indeterminate, difference between measurements in one frame of reference from those in another are equivalent to the rotation of axes in a four-dimensional manifold. In contemporary physics space and time have been united, and space–time is itself regarded as the metrical field, whose structure depends upon the distribution of matter and energy, themselves but equivalent forms of an identical substance. Matter, so far from being resoluble into hard impenetrable atoms, is now regarded as a singularity— "a pleat or chimney" (to use Eddington's terms)—in space–time, and all forces have been transformed into space–time curvature converting dynamics into geometry. In short, the universe, understood relativistically, is a single, seamless space–time whole.

Einstein and Infeld declared that materialism and mechanism have now been abandoned. Field has taken precedence over particle in importance, and matter is no longer viewed as merely particulate. Matter has been revealed as equivalent to energy, and the inseparability of the two concepts has been firmly established. The notion of field is holistic, for every variation in the field is determined by the pattern of the whole and is necessarily and inseparably connected with every other, and the field in principle extends over the whole of space–time. In wholes such as this, whose parts determine one another mutually and are interrelated in an ordered system, all relations are internal to their terms, a condition firmly established by the theory of relativity. The fundamental presuppositions of Newtonian science have thus been cancelled out and an entirely different world picture introduced.

Particles in certain circumstances (e.g., electrons projected through neighboring apertures in a screen) behave as waves, and waves have turned out to be particulate (photons). Particles, in fact, have come to be conceived as wave packets, as superposed waves which cancel one another out everywhere except in a sharply restricted volume. Electrons orbiting the nucleus of an atom are held to be standing waves (similar to the vibrations round the lip of a sounding bell) rather than separable particles in motion. Every type of particle is associated with a field, and every field with one or more kinds of particle. In short, atomism has given way to the conception of a whole (field), the structure of which determines the nature and behavior of the parts (quantum events). In quantum physics certain parameters are found to be complementary, so that neither can be measured precisely without rendering the other completely indeterminate. All that can be definitely determined is a probability amplitude within the system as a whole, which "collapses" only with measurement. In contemporary physics such holism is pervasive.

Heisenberg's principle of indeterminacy was for long a source of doubt and trouble to Einstein who maintained that "God did not play at dice." With Podolsky and Rosen he devised a thought experiment to prove that quantum descriptions were incomplete. This so-called EPR paradox gave rise to the hypothesis that there might be hidden variables governing the quantities that conditions of experiment prevented scientists from determining exactly. More recently Bell's theorem has demonstrated that such hidden variables cannot exist. Meanwhile, Bohm and Aharanov devised and carried out a modified form of the EPR experiment, confirming quantum theory predictions. Henry Stapp, arguing on the basis of these developments, has proved that there must be faster-than-light influences that are not signals (for signals cannot travel faster than light) connecting events at a distance. The states and motions of quantum systems have thus proved to be inseparably interconnected (nonlocal). Quasi-crystalline structures have been discovered exhibiting hitherto prohibited icosahedral symmetry, the assembly of which can only be nonlocal, as the state of the atoms at a distance from the assembly point affects the way the interfaces are aligned. In brief, the physical universe must be a single unified whole of interconnected events and overlapping fields.

Chaos theory has reinforced the demand to conceive the universe as one whole. In complex dynamic systems the priority of the structure of the whole (governed by fractal geometry) has been recognized as determining the nature of the parts, and the sensitivity of initial conditions in producing effects on the subsequent outcome has been shown to be so delicate that the minutest difference in the former can produce enormous consequences in the latter at remote distances.

Beginning with Edward Lorenz's discovery of the "butterfly effect" and strange attractors—that is, dynamic structures toward which all the motions in a complex dynamic system tend at specific degrees of energy flow—a number of mathematicians and physicists have discovered that turbulence is regulated by patterned forces that follow regular principles and form self-representative fractal systems that underlie all manner of natural and behavioral processes, from the great red spot on Jupiter (an example of stability within chaos) to physiological and morphogenetic cycles, anatomical structures, the forms of plants and animals (diatoms and shell formations), and even price fluctuations in market economies. Thresholds at which phase transformations occur prove to be the boundaries between the pulls of strange attractors. These boundaries turn out to be fractal forms corresponding to the shapes of natural objects such as snow flakes, fern leaves, foraminifera, and many others. The upshot of this area of research is that system is prior to detail and that reductionism is fruitless as the means of explanation.

Further, Heisenberg's principle of indeterminacy has persuaded his followers that quantum systems (i.e., microscopic physical structures) do not exist (or exist only potentially) until actually measured and observed. The Schroedinger-cat paradox extended this indeterminacy to macroscopic structures, and John von Neumann has shown that a psi function can be formulated that includes the whole of the measuring apparatus. The probability amplitude, Eugene Wigner has contended, will collapse only when the information enters the mind of the observer. Whereas in the Newtonian paradigm the observer was firmly excluded from what was observed, that the observer and observed are mutually inseparably implicated has now been established, not only by relativity theory, but even more so by quantum theory.

The search for a unified field uniting all known forces was first adumbrated by Schroedinger, Weyl, and Kaluza, and pursued by Einstein (without success) to the end of his life. The attempt to unify relativity and quantum theories spawned infinities that defeated the mathematics, until Richard Feynman discovered a way to "renormalize" the relevant equations. Since then theorists have resumed the quest for the unified field and in recent decades have come within fair prospect of reaching their aim. Quantum field, S-Matrix, Grand Unified, and Super-String theories bid fair finally to unify the entire field of the physical world and to be able to express it in the sort of fundamental mathematical equation adumbrated by Heisenberg as

some quantized nonlinear wave equation for a wave-field of operators that simply represents matter, not any specified kind of waves or particles. This wave equation will probably be equivalent to rather complicated sets of integral equations, which have "Eigenvalues" and "Eigensolutions," as physicists call it. These Eigensolutions will finally represent the elementary particles.[3]

Contemporary physicists have not yet found such an equation, but they have been able to unify three of the four primary forces and may soon be able to find a formula which includes the last (gravity) as well. The theory which promises this consummation is the Super-String Theory, which takes account of all the essential symmetries in such a way that, to be self-consistent and avoid all anomalies and divergences, there can be only one viable theory of the universe.

The clear message of contemporary physics is that the universe is a single undissectable whole, in which every entity and every event is dependent upon and inseparably connected with every other. In the third chapter of *Cosmos and Anthropos*, I have set out as well as I know how the nature and course of this revolution in the conception of the physical universe; and, in succeeding chapters, I have tried to show how it has spread to include a similarly holistic approach to biology, culminating in the recognition of the unity of the biosphere, which has come to be seen by biologists as a single indivisible biological community, organically related to an environment as intricately adjusted to the requirements of life as living creatures are adapted to their surrounding habitat. Not merely is the living organism a unity in which the parts are shaped and function as required by the nature and self-maintenance of the whole (a fact recognized long ago by Aristotle), but the life of the individual organism is equally dependent on the activity and functioning of the other organisms interacting with it in its environment. This interdependence is what is called "the balance of nature," and it prevails throughout the biosphere, a constant testimony to the mutual interdependence of living species among themselves and with their inorganic surround.

Such interrelatedness of living forms among one another and with the surrounding environment has been vividly described by Marston Bates:

Here on the atoll of Ifaluk, the distinction between land and sea seemed to lose its biological meaning. We could find no logical way of subdividing the environment into a series of discrete biological communities and we came to the conclusion that the meaningful community included the whole atoll situation: land, reef, lagoon and immediately surrounding sea. Everything was all mixed up. The

people depended equally on the land and the sea for their food. The
hermit crabs that crawled everywhere over the atoll went to sea to
lay their eggs, as did the the coconut crabs and the land crabs; the sea
turtles came out to bury their eggs on land. The influence of the sea
was everywhere; it determined what plants and animals were living
on land, because all of these, to get there, had to have some method
of crossing the sea, unless they had been purposefully or acciden-
tally brought by man. . . .

. The biosphere, then, is essentially continuous in space, a single interwoven web of life covering the surface of our planet. But it is far
from being a uniform, monotonous web: it is woven into a motley
series of patterns and designs.

Lewis Thomas has given expression to a similar vision. He has
compared the entire earth with its atmospheric mantle to a
single living cell, with interconnections between creatures in
the Sargossa Sea and the flora and fauna of the Alps. J. E.
Lovelock has put forward what he calls the Gaia hypothesis, arguing that the earth as a whole is a single living entity maintaining itself and its environmental conditions for its own survival.
There are many physical features of the earth and its surround, he
contends, that cannot be sufficiently explained from purely
physicochemical causes, and some which are kept constant by
the activities of plants and animals. Rupert Sheldrake has proposed the introduction into biology of the field concept, which
is essentially that of holism. He maintains that there is a morphogenetic field determining the processes of generation and
ontogenesis. The fact that these theories are scouted by other
biologists does not detract from the evidence of a growing sense
among scientists of the need to think holistically.[4]
    Evolution of specific forms of life has been reduced by neo-
Darwinists to the process of chance variation and natural selection. The latter, however, is something of a misnomer, as
nature does not select. Those specific variations which put the
organism at a disadvantage in the competition for sustenance
or in the conditions of survival naturally fail to reproduce and
die out. "Natural selection" is thus simply the elimination of

the unfit. Neo-Darwinism then regards variations that have positive survival value as the result of pure chance. These are, however, so rare and so improbable that it would take more time than has elapsed since life has appeared on earth to produce by pure accident the enormously complex and delicately contrived structures of contemporary organisms, as well as the complicated behavior patterns (often involving several different individuals) that contribute to the survival of living species. Some other influence must have operated than mere chance in the course of evolution. Stuart Kauffman has put forward a cogent case in favor of the idea that there is at work a principle of self-organization which, in the first place unifies the organism as such, and in the second integrates mutations into its structure. This, he thinks, occurs at the phase transition between chaos and order; in other words, it would be at the boundary between specific strange attractors. It is not inconceivable that the same self-organizing principle adapts the organism to its environment, produces the organic interdependence of the members of biological communities and ultimately welds them all into one interrelated whole—the biosphere.

The outcome of evolution of this organic totality is intelligent life; and that has now come to be seen by some physical cosmologists as necessarily and inseparably involved in the entire holistic structure. This has been expressed as the Anthropic Cosmological Principle, variously enunciated by the cosmologists, not always satisfactorily from a philosophical point of view. Yet it has a significant philosophical implication. The physical, meteorological, and astronomical conditions prevailing on this earth (and conceivably on other planets elsewhere in the universe) are precisely and delicately adapted to the requirements of life, and the elements indispensable for living matter (carbon, oxygen, phosphorus, and the like) have been synthesized in the interior of main sequence stars, for the generation of which the whole history and evolution of the physical universe has been necessary. If, as contemporary theories tend to insist, this universe is the only one possible that preserves all the required symmetries, it would seem to follow that the emergence within it of intelligent life is necessarily involved. It is a universe that inevitably produces observers, on whom the actualization of its minutest details depend.

The Anthropic Cosmological Principle has been propounded in several different forms. Of these formulations some have seemed merely truisms, while others have been questioned and even ridiculed. The primary concern has been that what we observe of the physical universe is conditioned by the fact that we inhabit it and are here to observe it. It is obvious that what we observe must include conditions requisite for our own existence, and while some physicists consider this an important heuristic principle, others have considered that it imposes a regrettable limitation upon our knowledge. They have speculated that the universe we observe may not be the only one possible; there may be others unobservable by us and conceivably inimical to the evolution of intelligent life; or our own may be but one of a succession brought about by alternate expansion and contraction (produced by big bangs followed by big crunches), of which those we do not observe exclude conditions suitable for life. If our existence here and now has been brought about in some such fashion, it could be merely a chance occurrence, not at all inevitable in the nature of things.

In Chapter 1 of *Cosmos and Anthropos*, I argued that these speculations are incoherent, first because there can be only one universe, even if it includes many separate worlds, and our speculations about alternatives are as much subject to the Anthropic Principle as what we observe, because it is we who speculate and infer their possible existence from what we do observe. Second, the most advanced contemporary theories have come to the conclusion that no universe organized differently from the one we observe would be viable, as any other would violate symmetries essential to its existence.

The true significance of the Anthropic Principle is that it follows from the discovery of the undissectable unity of the universe and the indispensable involvement of the observer in what is observed, as has been demanded by both relativity and quantum theory. This intrinsic involvement of humanity in the physical nature of the world runs directly counter to the presupposition of Newtonian science for which nature was closed to mind, and life was a fortuitous occurrence in a mechanical structure, itself no less mechanical than its physical surround.

A veritable galaxy of distinguished physicists has testified to the holism of this contemporary change in the scientific conception of the world. Max Planck writes,

Modern Physics has taught us that the nature of any system cannot be discovered by dividing it into its component parts and studying each part by itself, since such a method often implies the loss of important properties of the system. We must keep our attention fixed on the whole and on the interconnection between the parts.[5]

P. W. Bridgeman says,

We do not have a simple event A causally connected with a simple event B, but the whole background of the system in which the events occur is included in the concept and is a vital part of it. . . . The causal concept is therefore a relative one, in that it involves the whole system in which the events take place.[6]

Louis de Broglie states the same message:

The notion of the individuality of the particles is seen to grow more and more dim as the individuality of the system more strongly asserts itself. It therefore seems that the individual and the system are somewhat complementary idealizations.[7]

D. W. Sciama likewise is assured that

the universe is not a collection of independent objects. Its different regions strongly influence one another.[8]

Sir Arthur Eddington, writing of the expanding universe, remarks,

I only want to make vivid the wide inter-relatedness of things.[9]

E. A. Milne set out to deduce diverse phenomenal laws from a few general principles, and in doing so he showed that the motion of a free particle is dependent on "the rest of the universe.[10] Werner Heisenberg wrote,

The world thus appears as a complicated tissue of events, in which connections of different kinds alternate or overlap or combine and thereby determine the texture of the whole.[11]

David Bohm, describing what he calls the implicate order of the physical world, declares,

A centrally relevant change in descriptive order required in the quantum theory is thus the dropping of the notion of analysis of the world into relatively autonomous parts, separately existent but in interaction. Rather, the primary emphasis is now on *undivided wholeness*, in which the observing instrument is not separable from what is observed.[12]

Comparing the world view of contemporary physics with that of eastern mysticism, Fritjof Capra (writing in *The Tao of Physics*) is just as emphatic:

The basic oneness of the universe is . . . one of the most important revelations of modern physics. It becomes apparent at the atomic level and manifests itself more and more as one penetrates deeper into matter, down to the realm of subatomic particles. . . . As we study the various models of subatomic physics we shall see that they express again and again, in different ways, the same insight—that the constituents of matter and the basic phenomena involving them are all interconnected, interrelated and interdependent; that they cannot be understood as isolated entities, but only as integrated parts of the whole.[13]

The same idea is implied in the work and comments of many other scientists, including Einstein, Infeld, Schroedinger, Stapp, Barrow, and Tipler, to all of whom I have referred in previous publications.[14] Without serious question, therefore, one may claim that a new scientific paradigm has been established and that it is essentially and inclusively holistic. But, as we have seen, it has not penetrated beyond the physical and biological sciences. The thinking of the majority of philosophers and the general public is still enmeshed in the schema of concepts that prevailed in the nineteenth century.

As these concepts have brought contemporary civilization to the brink of self-destruction, there is an urgent need for change, and as this change has already occurred in the sciences, it should have repercussions elsewhere. What are its implications for philosophy, ethics, politics, and religion, and how should it affect our way of living?

It should first be noted that the universe as envisaged by contemporary science deploys itself spatiotemporally as a scale of forms progressively increasing in complexity and integrity. Each phase is a provisional whole, and successively they represent an advance in coherent integrity. At the base is space–time, which, through indeterminacy of activity at the micro-level, generates virtual particles in the vacuum and so becomes a field of energy, concentrating into matter in the form of elementary particles. These are ranged as mesons, leptons, and hadrons, and combine into atoms, each with its characteristic structure determined by Pauli's Exclusion Principle, the application of which results in organization. Every atom is thus a whole in itself, yet each has a tendency to combine with others to create new and more complex wholes. Atoms combine as molecules, similarly organized, and these cohere in the leptocosm of crystals, which are larger and even more coherent wholes. Crystals grow and combine in ways that foreshadow living activity, and liquid crystals prefigure protoplasm in which they are sometimes constituent. Macromolecules (themselves complex wholes) proliferate into polymeric chains to constitute organic substances, proteins and chromosomes, whose properties depend on the patterns of their combination. These combine further to constitute living cells. Some of these maintain themselves as protozoa; others segment to form multicellular organisms which ramify in sequential genera, increasing in complexity and versatility, until eventually intelligent species evolve.

Observe further that the forms in the scale overlap and each successive whole includes and transforms the character and activity of its predecessors. The scale progresses by what Teilhard de Chardin called *enroulement sur soi meme* (enfoldment upon itself). They are moreover at the same time mutually opposed and complementary, so that what at one stage are mutually contradictory, at the next stage are reconciled and united in the superseding whole. For example, the self-enfoldment of space–time is curvature, which in relativity theory is equivalent to force. Energy fields overlap to create wave packets, elementary particles of matter, an overlap of wave motion and particle motion. The four-dimensional manifold of space–time and the spreading light wave that physicists have identified are at once opposed, yet complementary concepts. Likewise, energy and matter are opposites while they are equivalent and complementary. Wave and particle are mutually contradictory in that the former spreads in concentric spheres, while the latter moves in linear trajectories. Yet they overlap—the particle is the self-enfoldment of wave motions, and they unite to constitute a new kind of subsidiary whole, in forms which again are ranged in a scale (according to Gell-Mann and Nee'man). Quarks overlap to form mesons and baryons; overlapping fields corresponding to neutrons and protons coalesce to become atomic nuclei; these again are positively charged electronically in opposition to electrons, with which they combine into atoms where the contradictory charges are reconciled and united. Atoms sharing electrons overlap and combine to form molecules, and so on to the macromolecules of proteins and nucleic acids, whose properties depend on their self-enfoldment and the overlap of their chemical bonds.

This dialectical relationship between the forms becomes more complex as the scale proceeds through the Mendlejev table of elements and the succession of molecular and crystalline structures up to living cells. The biosphere differentiates itself into a similar scale: Protozoa are opposite forms to metazoa, yet in the latter the character of both are combined and overlap. In ontogenesis the embryo begins as a single cell, the union of two opposed gametes, nucleus and cytoplasm being opposed but complementary constituents of the egg. The process continues by cell division, the blastula stage corresponding to primitive metazoa in which the cells are equipotential for further development, whereas in the stage succeeding gastrulation they become specialized to form diverse organs, distinct and variously opposed to one another in embryogenesis. Equipotentiality and specialization are opposed capacities; but finally the diverse organs are all inseparably and intricately united into one organic whole. The entire scale is vast, complex, and ramifying, and we have given only some examples here of its dialectical structure. This emerges eventually as the dialectic of human thought, the organizing principle of primitive sentience, of the experience of an ordered world, of intelligent action, and of scientific theorizing itself.

The physical universe has turned out to be a single whole structured as a scale of subordinate wholes from stardust to stars, planets, and galaxies, from elementary particles, through atoms and molecules to crystals. Likewise, the biosphere is constituted as a complex scale of ramifying genera and species increasing in complexity and versatility. Moreover, it turns out to be a single biotic community made up of subordinate communities, each an integrated symbiosis, and all of them populated by organisms, each of which is a whole of systematically related parts, themselves subsidiary wholes that are mutually ends and means subserving the self-maintenance of the organic system. These parts range from macromolecules of amino and nucleic acids to complex and varied proteins and chromosomes, from the metabolism of protozoa to the complex homeostatic chemical cycles subserving the physiology of multicellular organisms and their interdependent organs and vital functions. The entire series at any point may be resolved in greater detail to disclose subseries with a similar structure, so that it is not surprising Benoit Mandelbrot should have suggested that the geometry of nature is fractal.

All this has important philosophical, ethical, and theological implications that have not yet been sufficiently appreciated. The presuppositions of Newtonian science have been totally subverted and should no longer dominate our thinking in other intellectual spheres as they still do. Whereas the Copernican revolution was reputed to have reduced the importance of man in the cosmos by removing the earth from the center, the Anthropic Cosmological Principle restores human intelligence to a centrally significant place in the universe as directive of its evolution and even essential to its very being; and it reintroduces teleology (understood as the direction of participating processes by the structure of the whole system) into cosmological physics as a scientifically tolerable concept.

If the implications of this scientific revolution and the new paradigm it introduces are taken seriously, holism should be the dominating concept in all our thinking. In considering the diverse problems and crises that have arisen out of practices inspired by the Newtonian paradigm, it is now essential to think globally. Atomism, individualism, separatism, and reductionism have become obsolete, are no longer tolerable, and must be given up. This does not mean that analysis is useless, or that the examination of detail is unnecessary, but it does mean that reduction to least parts and examination of these will not by itself afford explanation of the structure of the whole they constitute. And as this structure is the directing influence and the determinant of the nature of the parts disclosed, the explanatory principle has to be the organization of the complete relational complex. In short, explanation must be teleological, for the proper import of teleology is the domination and direction of the part by the whole. Further, the parts discovered are to be treated as provisional wholes in their own right, participant in and contributory to more complex and more highly integrated wholes. Such holistic thinking would make an incisive and far-reaching difference to both theory and practice in every field of human interest and activity, the consequences of which we must next investigate.

NOTES

1. Max Planck, *The Universe in the Light of Modern Physics* (London: Allen and Unwin, 1937), p. 17.

2. A. Einstein and L. Infeld, *The Evolution of Modern Physics* (New York: Simon and Schuster, 1954), p. 158.

3. Werner Heisenberg, *Physics and Philosophy* (New York: Harper, 1958), p. 72.

4. Cf. Marston Bates, *The Forest and the Sea* (New York: Random House, 1960), p. 31; Lewis Thomas, *The Lives of a Cell* (New York: Viking, 1974), pp. 5, 41, 145–148; James Lovelock, *Gaia: A New Look at Life on Earth* (Oxford: Oxford University Press, 1979); Rupert Sheldrake, *A New Science of Life: The Hypothesis of Formative Causation* (London: HarperCollins, Palladin, 1987), p. 26.

5. Max Planck, *The Philosophy of Physics* (London: Allen and Unwin, 1936), p. 33. Also idem, *The Universe in the Light of Modern Physics*, p. 25: "It is impossible to obtain an adequate version of the laws for which we are looking unless the physical system is regarded as a whole."

6. P. W. Bridgeman, *The Logic of Physics* (New York: Macmillan, 1954), p. 83.

7. Louis de Broglie, *The Revolution in Physics*, trans. R. W. Wiemeyer (London: Routledge and Kegan Paul, 1954), p. 281.

8. D. W. Sciama, *The Unity of the Universe* (New York: Doubleday, 1961), p. 69.

9. Arthur Eddington, *The Expanding Universe* (Cambridge: Cambridge University Press, 1933), p. 120.

10. E. A. Milne, *Proceedings of the Royal Society of Edinburgh*, sec. A, vol. 62 (1943–1944), p. 11.

11. Heisenberg, *Physics and Philosophy*, p. 96.

12. David Bohm, *Wholeness and the Implicate Order* (London: Routledge and Kegan Paul, 1980), p. 134.

13. Fritjof Capra, *The Tao of Physics* (London: Fontana, 1985), p. 142.

14. Cf. E. E. Harris, *The Foundations of Metaphysics in Science* (London: Allen and Unwin, 1965; reprint, Lanham, Md.: University Press of America, 1983; Atlantic Highlands, N.J.: Humanities Press, 1993), pp. 136ff; idem, *Cosmos and Anthropos* (Atlantic Highlands, N.J.: Humanities Press, 1991), chapter 3 et seq.

6

Circumspective Transformation

LOGIC AND METAPHYSICS

I have represented the paradigm change as occurring with the work of Planck and Einstein, but Collingwood pushed it farther back and attributed it to Darwin. The science of the Renaissance, he argued, conceived the universe as a machine, but machines wear out—they do not evolve. Collingwood considered that the conception of evolution, when raised to the philosophical perspective (as it was for instance by Herbert Spencer and later by Bergson), introduced a new conceptual scheme.

The idea of evolution as it is entertained by contemporary neo-Darwinists would belie this interpretation because it sticks to a mechanistic understanding of chance mutation and natural selection. But I have tried to show that this mechanistic view of evolution is unsatisfactory and untenable. To the extent that we recognize evolution as implying holism of some sort, Collingwood's historical interpretation is acceptable. Insofar as adaptation is brought about by a drive within the organism to greater coherence, and inasmuch as it evinces an integration of organism and environment, it does involve holism; and indeed an organismic trend in biology has become apparent in the latter half of the twentieth century. Traces of this tendency were already evident earlier in the work of Hans Driesch, who postulated an "entelechy" to explain the homeostasis and equipotential capabilities of living cells and processes, expressly identifying the entelechy with the unity of the organism.

Bergson raised the conception to the philosophical level by making evolution a cosmological principle and reducing physical reality to a construct of the human intellect, a mere appearance enabling the living being to act in the circumstances in which it finds itself at a given moment. The reality behind this appearance, Bergson averred, is the "de-tention" of the life force, producing a sort of detritus deposited in the course of its evolutionary progress.

Another biologist, Lloyd-Morgan, pointed out that different and successively more complex wholes give rise to new emergent properties, and this idea, along with that of evolution, was exploited by Samuel Alexander to develop a complete new ontology. Alexander made a manful attempt to incorporate the results of the new physics into his theory of space–time, but with only questionable success.

These philosophers, however, attempted only the first, somewhat halting interpretation of the current paradigm. The more significant change came at the turn of the century, in physics rather than in biology, and the implicit holism of the new theories was much more far reaching. It was the mathematician A. N. Whitehead who produced a more significant metaphysical interpretation of the new physical theories. He contended that every actual entity was an event in a creative process of unification reflecting the unity of the entire universe. Of his doctrine of process in which each actual entity prehends every other, it would be true to say that the whole was in every part and every part contributed essentially to the whole.

The order and regulation of the whole that contemporary physics has revealed is determined by the universal organizing principle, formulated by the physicist as the mathematical equation envisaged by Heisenberg, from the "Eigensolutions" of which all forces and particles would be derivable. Philosophically this is what has been called the concrete universal, the organizing principle of a system or whole whose structure governs the nature of its parts and the process of their generation. Bernard Bosanquet strove to give an account of this universal, as system, but it was left to Collingwood to do so more successfully in his *Essay on Philosophical Method*. There he demonstrated that the philosophical universal was specified as a scale of overlapping forms related as opposites, as distinct species of the universal, and as a graded series of exemplifications of the generic essence, progressively increasing in degree of adequacy to its integral wholeness. This was at once a logical account of the nature of the philosophical concept and a metaphysical theory of its concrete manifestation.

At this point, however, the speculative development was arrested, blocked by the resurgence of Empiricism in the form of Logical Positivism, which reverted to the earlier paradigm and reasserted the position of Hume, rejecting all metaphysics as meaningless and restricting all philosophy to logic and linguistic analysis.

On the contrary, the remit of philosophy should be to examine the logic of system and the metaphysical nature of wholes, such as contemporary science postulates. It should be recognized that every such whole must be differentiated. It must be a unity of differences and can be neither punctiform, without structure, nor empty extension. In fact the latter is really inconceivable, for extension itself involves distinctions of position and direction. It must have dimensions—a nil-dimensional manifold is a nonentity. Again, to constitute a whole, the parts must be integrated according to some principle of structure, otherwise they would be no more than a haphazard congeries without unity or coherence. A whole then is not whole unless it is both differentiated, structured, and complete. The internality of relations entails the implication of the organizing principle in every part, determining the structure of the whole. Accordingly, no part can be self-sustaining without the rest, and each by its own nature posits the whole to which it belongs. Every partial existent therefore guarantees the reality of the totality.

Because each part is an essential factor in the whole, each has its own specific nature, and owing to its complementarity with all the other parts, each will be itself a provisional whole. It is provisional because it requires complementation by its other in order to be what in fact it is. Consequently every partial entity entails a tension between what it is and what it is not (its other), a tension that generates in it unrest and embues it with a nisus to more comprehensive wholeness in combination with that other. In this combination, the former opposites are not lost, although their mutual complementation abolishes the contradiction formerly apparent; in the new whole each is preserved but transformed into an aspect or moment of the whole. Their mutual relation is thus dialectical, and the dialectic is a spur or nisus to the whole. Accordingly, the principle of structure pervading the whole is never static, but is a dynamic principle of activity and progression. Evolution, in short, is implicit in the very nature of holism.

The result is that the universal principle of organization specifies itself in a series of overlapping forms, increasing in their degree of adequacy to the coherence and comprehensiveness of the overall totality, each contrasting with and in some sense contradictory of the others, yet also complementary to them, as they are to it. Because the organizing principle universal to the whole is implicit in each provisional phase, each is a distinct specific manifestation of this universal. Likewise, because each stands in contrast with the others, they relate to one another as opposites; and as each in its deficiency tends to unite with its immediate opposite to form a more adequate whole, they constitute a scale of degrees. Collingwood's doctrine of overlapping classes and the self-specification of the philosophical universal as a scale of forms is thus corroborated.

This in outline is the logic of system, or wholeness (for every system is a whole, and every whole is a system). It is also a theory of the ontology of the cosmos, and empirical science reveals a structure in nature corresponding to that required by this logic. For example, the nucleus of a hydrogen atom, under certain conditions, exists as an ion, but it harbors a tendency or nisus to combine with an electron to which it is in opposition, to form the atom. That again has a nisus to combine with a second atom to form a molecule (H_2), and that has chemical affinities to other atoms to form other elements and combinations such as helium and water (H_2O). The same general principle pervades the entire gamut of natural forms, becoming more complex as it proceeds and accordingly more difficult to penetrate and make theoretically apparent. We cannot, for instance, give a similar account of the relation between protozoa and metazoa, although it is fairly evident that some similar relationship lies behind the process of evolution from one to the other. Both the philosophical analysis and the scientific exemplification vindicate Collingwood's insights into the structure of what he called the philosophical universal, and they give testimony to the transition from the Renaissance to the modern paradigm.

The implications of the new scientific paradigm for philosophy then are first a logic which is not purely extensional (not one that can apply only to aggregates, or sets, in which relations are external), but dialectical, appropriate to wholes (in which relations are internal). Extensional logic will be supplemented by dialectical logic—in fact, dialectical logic, by recognizing that aggregates (to which formal logic applies) are low-level forms of whole (for the concept of a whole is itself specified as a scale of forms), can find a place within its own system for formal (or extensional) logic. Second, explanatory procedures cannot be simply reductionist—that is, they must not seek to explain the whole in terms of the parts in their separation, but, while distinguishing the specific differences in the minutest detail, will seek the explanation of the parts and their behavior in the structure of the whole. They will therefore be teleological. This does not mean that natural processes are to be regarded as purposive, but purposive behavior will be recognized as a special case, at a relatively high level, of the universal tendency toward holistic completeness. In the third place, contemporary ontology should likewise be dialectical, finding the specification of the organizing principle that governs the real in a scale of overlapping, graded forms that are nevertheless distinct, opposed, and mutually complementary, which

progress toward and together constitute a complete and perfect whole that is the ultimate cause (both efficient and final)
and the ultimate explanation of everything. The whole accordingly is the ultimate criterion of truth, and relativism is ruled
out of court.

The scale of forms in which the natural world specifies itself
does not end with the merely physiological. Physiology at a
crucial stage of development involves sentience, and sentience
is organized by attention into the cognition of objects, ordered
further as the perceptual world. At a subsequent stage of development, this consciousness blossoms into self-reflection and
rational self-awareness. The products of this phase of development form what has been called the noosphere, which is not
separate from but continuous with the gamut of nature.

Self-conscious awareness is moreover implicit in the very
notion of a systematic whole, because such a whole is a relational complex and relations are fully realized only when they
are cognized as such. The relations between the elements of a
system are merely implicit as long as the system is only physical or even organic. Elementary particles interact as their interrelations require, but in so doing they do not make the
relationship explicit (it becomes so only when indicated by the
scientist); likewise chemical affinity is the result of a special
relation between chemical elements, but it does not of itself
specify the relationship, which only a chemist can define. The
same is true of biological relations, they serve to maintain the
life of the organism, but they operate unconsciously and become explicit only when they are cognized, either directly by
that organism itself, or by the investigating biologist. Accordingly, to become fully explicit and actual, a relational complex
must become aware of itself—it must become the intentional
object of a conscious subject who is not confined to any term of
the relationship, but must transcend the complex as a whole.
So the quantum physicist contends that a quantum system is at
best potential and does not become actual until its parameters
are experimentally measured, and the probability amplitude
collapses only when it is observed by the experimenter. That
the relational complex becomes fully actual only when it is

brought to consciousness is the logical basis of the Anthropic Cosmological Principle, and has far-reaching implications. The physical cosmologist has established that the universe is an indivisible whole, a complete interconnected relational complex; and it is now apparent that such a whole (1) must be complete (otherwise it could not be whole), and (2) that it becomes completely actualized only if and when it is cognized by a conscious and intelligent observer. Intelligent life is therefore necessarily involved in the physical world.

ANTICIPATIONS OF THE PARADIGM

While, on the one hand, twentieth-century Empiricism was a throwback to the eighteenth-century paradigm, there were strands of thought in the preceding epoch which were ahead of the times, anticipating in various respects the discoveries of the present day. Spinoza's inspired insight led him to realize that thought and extension (mind and matter) could not be mutually independent but must basically be identical, and that the structure of both was hierarchical in gradations of complexity which, as it increased, produced entities progressively each more self-sustaining.[1] "The face of the whole universe" he maintained was a single individual whole and was substantially identical with the infinite idea of God. In effect, he also anticipated the Anthropic Principle, asserting that everything is beminded (*animatus*) and that the human mind is but the concomitant idea of a complex, self-sustaining body more capable than that of less complex bodies of clear and distinct (i.e., rational) perception. Similarly, Leibniz conceived the physical world as a system of *phenomena bene fundata* based on an infinite system of interconnected monads, each reflecting the whole universe in its "perceptions"—a universe that was therefore a self-representative system of internally related systems interconnected by a preestablished harmony.[2]

Kant realized that the world could be cognized only as a single whole by a single transcendental unity of apperception, and that teleology described an organic unity in which the whole determined the parts and the parts were mutually ends and means.[3] So Schelling and Hegel were led to decide that reality was a unitary whole, differentiating itself dialectically as a scale of forms through which nature brought itself to consciousness and culminated in an Absolute Mind (*Geist*).[4] These prophetic insights were directed in the first instance by reflection upon the nature of the knowing subject, in the effort to solve the problem of knowledge as posed by the Copernican paradigm. They were prompted by an awareness of the philosophical difficulties presented by that paradigm and the need for change. The cognizant *ego*, as Kant realized, by its spontaneous synthesis of sentient experience, orders its objects into a structured world in accordance with organizing principles derived a priori from its own transcendental unity. Consequently, the experienced world must be a coherent whole:

> The law of reason which requires us to seek unity is necessary, because without it we should have no reason at all, and without that no coherent use of the understanding (*zusammenhängenden Verstandesgebrauch*), and in the absence of this no adequate criterion of empirical truth. With this last in view, therefore, we must presuppose the systematic unity of Nature throughout as objectively valid and necessary. (*Critique of Pure Reason*, A 651, B 679)

Such premonitions in a former age of later developments should not surprise us. We noticed before that heliocentrism was anticipated by Aristarchus and the theory of impetus led the way to the conception of inertia. Mechanism and teleology (as I have demonstrated in Chapter 13 of *The Foundations of Metaphysics in Science*) are not mutually incompatible, as is evident from the holism of every machine. In fact, if holism is true it should apply to the history of science and philosophy as well as to their subject matter. Both seek coherence, and the whole, that is, the truth, will be immanent in all its specific forms, and just as later phases sublate and preserve earlier ones in a new form, so earlier phases will foreshadow the later. The dialectical logic of scientific revolutions is not simply a disjointed sequence; there is also a continuous thread connecting the opposites, which are in principle complementary.[5]

MORAL PHILOSOPHY

At the level of the noosphere, the dialectical relationships are highly complex, for self-reflection ensures that each phase in some way mirrors and represents the whole, while it is nevertheless but one phase in the dialectical process. Thus sentience registers the whole of nature through the unrestricted flow of influences from the environment upon the living organism. Primitive sentience is itself a whole of undiscriminated feeling which is distinguished into the various sensuous modes by attention, raising it to the level of perception. The organization of sentience into sense perception presents to cognition a unified world, the world of common sense, in what Husserl termed the natural attitude. This world is the same as the physicobiological world, now come to consciousness through the organization of bodily feeling, which is the registration within the organism of all the incident influences from its environment. Reflective thinking next produces science and philosophy as well as art and religion, and these are mutually related as degrees in a scale of figurative and explicit representation of the whole (or truth), each comprehending that whole and including the others within its own genre. Human reason emerges at this stage in the scale of forms as the organizing and uniting principle which has all along been active throughout the process of development. Social and moral action are forms of self-reflective (rational) conduct correlated with these noetic attainments and objectifying what in them is subjective experience.

The account we give of human society and conduct is therefore bound up with our understanding of nature as a whole, for they are the continuation of the scale of forms in which the natural whole is specified. Hence, what we conceive as morally obligatory will be what completes and perfects a whole of personal and social action within the context of nature, and so "what ought to be" will be inseparable from "what is." Ethical relativism is as little tenable as epistemological, and contemporary ethics should take on a new character.

Morality is an outcome of the development of self-consciousness—the essence of human reason, which has now transpired as the principle of organization that has been at work through-out the scale of natural forms, raised to the level of self-con-sciousness. As such, reason is the source and agency of order and unification. When it manifests itself in the life of human beings as social order, morality arises as a necessary aspect of social conduct.

Human action is directed by appetitive drives which, at the self-conscious level, become desires. At times, however, these come into mutual conflict, both in the individual person and between different persons when they find themselves in com-petition. In order to attain full satisfaction, not just of particu-lar desires, but of the self as a whole, reason has to organize desires and impulses, restraining some and enjoining others. This is the basis and source of moral obligation, the social equivalent of which in principle translates into political obli-gation. Its ultimate aim is a coherence and wholeness of life, the implications of which, when fully unfolded under condi-tions prevailing today in which common interests include the preservation of the planetary ecology and the maintenance of world peace, disclose the demand for a universal community of mankind and a concomitant universality of moral principle. Rela-tivity to the group and its culture proves to be merely provisional and finds its proper fulfillment only in the universal standard com-mon to all human beings. To work out this implication in detail would take us beyond the limits of the present discussion, but what is set out below will make it clear that nothing less than a world community, pursuing a good common for the whole of humanity can in future save us from apocalypse.

Now that the paradigm change has come in due time, we need to think globally in every respect. It has to be recognized that personality, while remaining of primal importance, is in-separable from community life. The social whole is a commu-nity of individuals, and each personality is a social product reflecting the structure of the culture in which he or she par-ticipates. Further, the local community cannot be separated from the national (ethnic) community, nor can that again be divorced from the life of the whole human race in its dependence on the global ecosystem.

Civilized living implies a rationally ordered community, and rationality and rational conduct is the outcome of a continuous evolution in nature, linking humanity inseparably with the natural whole. Consequently there can be no separation of fact from value; the latter is the result of evaluative judgment in a mind which is actually rational and is the product of a factual natural development. Human beings are by nature rational as they are by nature appetitive and passionate. What nature mandates is that they can achieve satisfaction only by rational organization of their natural drives and desires. This is therefore morally obligatory. What is seen to be good and right issues from the nature of persons, and that again is a product of the nature of the world. And as intelligent life is inherent in the order of the universe (in accordance with the Anthropic Principle), and as intelligent life involves social order and political organization, so reason and the law of nature prescribe the moral as well as the civil law as truly as they do the physical and the biological.

Ethics should now enter a new dimension. Certainly, there is still no denying the Cartesian *cogito*, but equally it is obvious that, as Husserl reminded us, *cogito* implies *cogitatum*, an intentional object. Subject and object, although opposites, are mutually complementary and in the last resort identical. The *ego* finds itself as being-in-the-world, a world that includes other persons without whom the *cogitata* molding one's own individuality would not be what they are.

Accordingly, individualism disregarding social intercourse is out of the question. Individual personality is a social product, and the interests and rights of the individual cannot be separated from those of the community. They are determined in large part by the social function of the subject performed in the service of the common good. As in physics observer and observed cannot be divorced from each other, so in sociology and politics individual and society are interdependent, and the social whole cannot be dissipated into a mere collection of persons. Moral obligation is intrinsic to civil living and needs no presumed contract for its validation.

Due to the influence of Plato and Aristotle, some moralists in the past have recognized this dependence, but the majority

have failed to respect it. Further, the common good is not confined to the local community or even to the nation. It must now
be seen to embrace all peoples the world over, for their welfare
is dependent on the conservation of the global environment,
which is not within the control of isolated individuals, and is
the responsibility of no one nation alone, but can be ensured
only by concerted action throughout the civilized world.

Accordingly, the good life, which was the end envisaged by
the ancients, is now dependent on a wider view than any polis
and must be interpreted in terms of universal survival and welfare embracing humanity as a whole. Rights and duties of the individual can be protected and ensured only in conjunction with
the recognition of appropriate rights not only of other persons,
but also of other species and even of inanimate factors of the
environment. Nature can no longer be regarded as a mere means
to be exploited for personal gain, but must be treated as an end
in itself. As we now have a duty to conserve the environment,
the environment will bear a corresponding right to respect.

POLITICS

Ethics and politics have never justifiably been separable, for
the common good of the whole is as much a political as a moral
end. Hence political obligation has a moral source which is not
just the obligation to fulfill a contract but is the universal duty
to serve one another and the good of the social whole. Few
would dispute the contention that political arrangements are
designed to achieve the welfare of the whole community, but
hitherto this welfare has been interpreted as the collective good
of individuals. Philosophers as well as politicians (e.g., Margaret Thatcher) have denied that there is any such thing as society, the reality being no more than a collection of persons. True
it is that no talk of common welfare has significance unless the
individual members of the group enjoy the benefits of cooperation; nevertheless if individuality itself is a social product
and is inseparably bound up with the person's function in society, the good of the individual can never be confined to the
person as such, because apart from his or her social relations it

is quite unattainable. The intermesh of social functions is the basis of all political organization, and unless the solidarity of communal action is firmly held in view in all political thinking, the social evils and problems that beset us today are hardly likely to be overcome.

Human beings, having a long infancy, are dependent for survival on the care of their parents. By nature therefore they live in family groups and tend to congregate in clan and tribal association. Being rational animals they organize their social relations, and to achieve a tolerable living standard, they divide labor and diversify functions. This implies duties owed by every citizen to all the rest who depend on him or her for some specialized service. Civilized life depends on the intermesh of social functions, ordered with a view to the common welfare of the community, and the individuality of each person is molded upon the special part that person plays within the social group. Moral duties are duties to others, dependent on the services performed and the recompense earned. They are summed up in the precept to "do unto others as you would wish them to do unto you."

Further, division of labor requires exchange of products and results in an economic system, which requires regulation; and organized human intercourse requires rules of conduct. Laws have to be enacted and promulgated, and delinquency has to be restrained. The result is a political system, with which no developed civilization can dispense. Sociopolitical law and custom are the framework of culture which is the material out of which morality and individual character are woven. Divine law itself is a social product, insofar as all religion, natural or revealed, is dependent upon human reflective self-consciousness and human ability to interpret the experience of life and the world in which it is lived; and this itself is the product, as it is also the source, of the rational ordering of human society and the educational system that it evolves. Without society there is no morality. As T. H. Green declared, "No man can make a conscience for himself; he needs a society to make it for him."

Individuality is thus a social product, and there can be no individual rights except those recognized by society, and none that pertain to persons by nature prior to the civil condition. The rights to life and exemption from "cruel and unusual punishment" may be universally claimed, but only when politically recognized do they become effective. In general, rights are the incidents of function, each person being entitled to whatever benefits are required for the efficient performance of his or her duty to the society of which either is a member.

It follows that the determinant of personality and its rights is the sociopolitical whole, and the result of thinking holistically in politics is far different from that of the individualism of the seventeenth century.

If the notion of unity and community pervaded all personal conduct, self-seeking and personal advantage could never be the paramount aim of action. Sleeze among politicians and political corruption would then become unthinkable, and the whole concept of democracy would be transformed. Parliamentarians would never represent merely vested interests rather than social needs. Lobbying would not be on behalf of commercial interests as much as for community services.

Today, however, the common good is not limited to the local group, because its welfare depends as much upon its commerce with other groups as that within itself. No nation nowadays can be wholly independent and self-contained, for each is affected by what is done and by what happens in every other; and as all are equally affected by environmental deterioration, the whole of humankind has become a single community, the common good of which is necessarily implicated in the good of every individual and every society. Contemporary politics as well as contemporary ethics should therefore be global, with consequences to which I shall return after a brief consideration of the changes needed in our thinking about economics.

ECONOMICS

World economics has already begun to take over from a "political economy" that concentrates on production and distribution within a single community. Not only have entrepreneurial ventures become transnational and national economies interdependent, but the methods of production and distribution are worldwide in their scope and are having effects on the ecosystem, the cost of which can no longer be ignored. Economic growth can no longer be sustained with impunity irrespective of the sustainability of the sources of energy and raw materials. The economic health and success of every country is dependent on that of all others, so the world economy has to be seen as a single system and must be treated as a whole. Further, the conception of profit must be transformed: It must be socialized rather than individualized. Production and supply have to be viewed as a cooperative enterprise rendering service to the community, rather than a venture undertaken for personal gain. Likewise labor is not to be exploited to ensure profits, but has to be employed in partnership with capital for a common good. Wealth has to be justly distributed if all are ultimately to reap the benefit of economic activity, for poverty is an inevitable drain on public funds and the health and vigor of the nation depend on the general level of welfare within the community. The cost of "welfare" in every society is thus consequential upon income differentials.

Estimations of wealth cannot in the future ignore the damage done to the environment by the methods of production and manufacture. Creation of manufactured capital has to be offset by the destruction of natural capital. The concepts of economic growth and development have to be redefined in terms of sustainability. No growth is indefinitely sustainable, and present needs have to be met without endangering the supply of future needs. Economic growth can only be truthfully assessed by taking account of population growth and the possibility of irreversible damage (what cannot be repaired) to the planetary ecology. Moreover, besides the needs of the present generation, those of future generations have also to be taken into consideration. Economic policies, both national and global, will henceforth have to be revised with these factors in mind so as to ensure that growth is limited to what the environment can sustain and to what does not irreversibly damage natural capital. In short, a global outlook has to be adopted, both with respect to populations and to the environment; human economic activity has to be seen as a whole, inseparable from the terrestrial whole within which it operates.

WORLD ORDER

Neglect of these considerations in the pursuit of ends inspired by the Newtonian pardigm has plunged humankind into a global crisis from which it can extricate itself only by thinking globally. The persistent worship of the fetish of national sovereign independence runs counter to this need and prevents what is currently called the international community from maintaining world peace. National sovereignty, however, came into being in the effort to achieve the common good of a cohesive society. Today, that good depends, in all cases, on the maintenance of world peace and the elimination of the threat of nuclear and other mass destruction. It depends crucially as well upon the conservation of the planetary ecosystem. For reasons that we explored in a previous chapter, these aims cannot be achieved as long as nations remain sovereign and independent; in fact, it is clear as daylight that no nation is in actuality any longer independent of others. Nor can the environment be protected by means of international agreements, because sovereign nations first of all give precedence to their national interests, and further because they can only be held to the observation of treaties by the use of military force, which defeats the object of any agreement intended to maintain the ecological balance of nature. In consequence, national sovereignty is no longer competent to serve the common good of the people over whom it rules, and the ethical justification of its power has withered away.

It follows that a political system must be devised that can effectively legislate for the common good of humanity as a whole. Such a system would have to be democratic and federal in form. First, to be in the true interests of the whole of humankind, no world government can be authoritarian. Second, to be acceptable, no world government can be forcibly or imperiously imposed. World government must therefore be democratic—it must be popularly elected and approved. It will be so only if the central authority confines itself to legislation affecting issues of universal common concern while it preserves the right of member national communities to autonomy. Ultimately, world government must be federal in form. It must be a unity of differences.

At the present time, because the customary way of thinking is molded by the Newtonian paradigm, few people give serious consideration to the establishment of world federation; but if, in accordance with the new scientific holism, global thinking were to prevail, that would be our immediate political goal; for it is the only condition under which global measures could be enacted to cope with the global problems which face humanity and threaten the survival of the species. Clearly, such all-embracing actions cannot be accomplished by the individual efforts of private persons and organizations. They must be universally enforced, and that requires global legislation and the global maintenance of a rule of law, which is possible only under a world government.

The international crises and tangles in which the nations nowadays continually become involved would not occur within a world federation. For example, the shocking situation that developed with respect to Kosovo need never have happened had there been a properly constituted world court competent to adjudicate between the Kosovar minority and the Serbian majority. Any attempt to use force or implement ethnic cleansing would have been out of the question, and anyone attempting such methods (Milosevic or whoever) would have been arrested by the federal police, prosecuted by the federal attorney general, brought before the world criminal court, and committed according to law. No military action would have been called for, and there would have been no massacres or mass expulsions, no destruction of villages, no refugees, and no unpunished violation of human rights.

RELIGION

Spinoza held that prophecy was revelation of the truth through the medium of the imagination, and Collingwood maintained that religion was artistic imagination asserted as the truth. For Hegel religion was the absolute Idea in the form of representation. Believers, on the other hand, have seldom acknowledged the fact that religious scriptures and doctrines are for the most part metaphorical and symbolic rather than a literal statement of the word of God. Without blatant incoherence they cannot all be taken literally as statements of fact. Such scriptures have to be recognized and interpreted as historical documents, the contents of which are colored by the character and beliefs of the authors, as well as the understanding of the intended readers. In whatever way religious dogma is viewed, however, Newtonian science always seemed to be in serious conflict with it.

The new paradigm, however, has completely transformed the situation created by its predecessor with respect to the relation of science to religion. In the first place, the conception of the universe as a whole, and the immanence of the whole in every part, point to a fresh view of the relation of the infinite to the finite, giving new insight into the conception of God and of the relation of the divine to the created. In the second place, the conflict between Darwinism and the Scriptures can now be resolved, because holism has recast the idea of natural selection, making it something more than the merely negative elimination of the unfit. Recognition of the organic wholeness of the organism and of the biocoenosis transforms the concept of evolution. What natural selection preserves is the more organically integrated, both intrinsically and within the ecosystem. It is impossible any longer to believe that mutations with survival value can be purely accidental. The improbability of their producing, by pure chance, in the time available since life first appeared on the planet, the highly complex and delicately integrated living systems that have already evolved is astronomically high. There is now impressive evidence of a principle of self-organization, coordinating genetic mutations and integrating them into the already existing organic system.[6] Evolution, therefore, is organismic and involves, as it is involved in, holism. In the organism it is the whole that determines the nature and behavior of the parts, and this direction by the whole also governs the integration into it of mutations, be they accidental or induced. Natural selection then presents no obstacle to the dialectic, which rather is its source and driving force, giving significance to the opposition as well as to the adaptation of organism to environment. When the full implications of the dialectic are worked out, moreover, they lead to a theistic conclusion.

Scientific theory can now be seen as compatible at least with a philosophical interpretation of religious teaching. The concretely universal, organizing principle of the unitary whole revealed by contemporary physics is obviously the creative principle of the universe. It is a principle of wholeness that diversifies itself into a scale of forms increasing in adequacy to the unifying drive of its own action until it emerges as self-conscious reason—rational personality. But the whole is essentially complete, and thus the finite rational personality, still imperfect in ourselves, cannot represent the final perfection, which must be real, for it is the principle of all being and the determining essence of every partial form. Without the whole there could be no parts. The total actualization of the principle, however, can be nothing less than a complete cognition of the entire relational complex. It can but be conceived as an absolute self-conscious mind encompassing the entire gamut of natural forms and transcending everything in space–time, everything finite. This answers precisely to the definition of God given by Anselm ("that than which a greater cannot be conceived") and is the only conception of deity worthy of worship. As the human mind transcends the body and all spatiotemporal relations in the act of cognition, the subject of which cannot be confined to any one of the related terms, so in its self-conscious grasp of its own self-specification, the divine mind transcends all finite forms as well as the entire gamut of natural phases that it encompasses.

As I tried to demonstrate in *Cosmos and Theos*, all the traditional proofs of the existence of God can be revalidated and reinstated on the basis of the scientific conception of the universe current at the present time, if it be recognized that a whole (such as physics envisages today) necessarily deploys itself dialectically in specific forms that are degrees of its own perfection. This leads to a metaphysical *Weltanschauung* that involves the Anthropic Principle—the necessary implication in the physical world of intelligent life. Intelligent life is, as Whitehead saw, a late phase in the concrescence of the universe, which as it proceeds culminates necessarily in the absolute perfection of the divine. Rational self-reflective intellect (what in the Book of Proverbs is called Wisdom) is created in the image of God (as the Bible proclaims). The universal creative principle of order—the *logos*, or (as otherwise translated) the Word—is prior, logically and ontologically, to all finite existence. It is with God before all created being, and it is God, as the opening of St. John's Gospel tells us. It is made flesh, in us and most eminently in Christ. The unification (atonement) of the finite and the infinite, sought by all genuine religion, is manifest in the identity of humanity and divinity in Christ. This is the supreme religious affirmation of the Anthropic Principle.

If religious teaching is taken, as it should be, as a metaphorical and symbolic representation of the truth, and the philosophical implications of current scientific theory are properly developed, they converge, and there is no longer any conflict between science and religion. Scientists no longer need to strain their consciences in the effort to live in two incompatible worlds at the same time, as they did when the Newtonian paradigm prevailed. One can be a believer in God and in religion and accept what science reveals without inconsistency.

TWENTY-FIRST CENTURY PROSPECT

There is little evidence that people in general, certainly, other than scientists, neither academics nor politicians, are yet thinking along these lines. The new scientific paradigm has by no means permeated the attitudes and ways of living of the mass of the people. This is hardly surprising when one considers that the Aristotelian paradigm was the fruit of three centuries of development and the Newtonian took at least two hundred years to become widely influential. But the situation of humanity today is unprecedented. Human activity, conducted under the influence of Newtonian thinking, and using the technology Newtonian science has engendered, is destroying the environmental conditions that support life on earth. Unless this tendency can be reversed, it is improbable that the Einsteinian scientific revolution will ever bear fruit in the social, moral, economic, and political practice that should naturally follow from it.

The problems that confront us are global and can only be resolved by global action, such as the new paradigm should inspire; and unless such global action is taken promptly the deterioration of the living environment will (if it has not already in some respects) become irreversible.

Some indication has already been given of the extent to which this deterioration of the environment has progressed. The protective ozone layer in the upper atmosphere, which shields living things from the harmful, often lethal, incidence of ultraviolet rays, is being destroyed by the excessive use of fluorocarbons. Unless this process can be reversed, cytoplankton, the basis of the entire food chain, is in danger of decimation, food crops are threatened with destruction, and human beings will be exposed to radiation, causing skin cancer and other diseases.

It has now been scientifically established that global warming is accelerating; nor can there be any doubt that human activities are contributing substantially to the change. Not only has the emission of greenhouse gases from the use of fossil fuels exceeded all admissible bounds, but the greenhouse effect is accelerated by the prodigious and unchecked destruction of the rain forests which is depriving the planet of the major source of atmospheric oxygen and one of its main means of absorbing CO_2. This destruction is annihilating thousands of living species, some of which are essential links in the food chain, and others which are medicinally valuable, and is depriving indigenous tribes of their established habitat and ecologically benign way of living. Moreover, the burning of forests to make room for agriculture exhausts the soil and, after a few years, leaves it arid, while reduction of vast areas of evaporation and transpiration reduces rainfall and contributes further to desertification.

Industrial waste, some of it extremely hazardous, is polluting, air, water, and soil, threatening or often outrightly destroy-

ing fish, birds, crops, and people. Marine oil spills and the practice of sluicing out oil tankers at sea destroys the breeding
grounds of plankton, kills marine life and seabirds, and does
widespsread damage, some of it unheeded and unrecorded.

The uncontrolled burgeoning of human population is making unsustainable demands upon terrestrial resources, further
destroying the environment by overcultivation of the soil, as
well as through the construction of roads and the spread of
built-up areas which prevent rain water from soaking into the
earth. This causes increasing floods and landslides in bad
weather, which global warming is making more frequent and
more severe. The increase of tourism is having similar deleterious effects. In consequence, natural disasters are becoming more
frequent and more destructive.

Genetic modification of crops, while it is intended to increase
yield and so better to supply the needs of the growing population, will probably lead to the extinction of the diverse natural
species from which those now cultivated originated, depriving us of the means on which we should rely to produce new
hybrids, if and when those we now cultivate become subject to
new diseases or the ravages of unsuspected insect pests. Moreover, the genetic modification of species may have side effects
which are as yet undiscovered.

In addition to all this, international relations are still being
conducted in consideration of a balance of power between sovereign states, ranged in opposing blocs. The collapse of the Soviet Union and the end of the Cold War has only temporarily
reduced international tensions. It has not eliminated nuclear
arsenals. Nothing prevents the possible rise of China as a new
superpower or the development of a new cold war. The arms
race has moved from the superpowers to the third world, and
the unchecked, and probably unpreventable, proliferation of
nuclear capacities sustains the menace of nuclear annihilation.

The danger of accidental nuclear disaster has been increased
by the decrepit condition of the Russian and Ukrainian nuclear
installations and the temptation to sell nuclear materials and
know-how to other countries in order to acquire hard currency.
Worse than this is the fact that the nuclear powers are being

overtaken by the menace of the computer millennium bug (the Y2K computer problem). Over a million computers in thousands of automated systems coded in seventy different computer languages involved in nuclear bomb, and power installations have to be deactivated before January 2000 if accidental firings and meltdowns are to be prevented. This may well prove to be impossible.

The reasons why current practices militate against the appropriate measures to counteract these menacing trends have already been given. Unless public thinking can catch up with the twentieth-century scientific paradigm, the end of civilization will be in sight before many years of the twenty-first century have passed—and not just Western civilization or modern civilization, but all prospect of future civilization as well, and possibly even the end of the human race, if not of all life on earth.

We must think globally. The aphorism, "Think globally and act locally," is all very well, but if local action is not widespread (or rather, universal) it will not avail. It is not even enough to realize the dangers and the necessary measures to counter them, unless those measures are incorporated in legislation that can be enforced upon everybody. Humanity must become a single unified community—that is, a society in which the common interest takes precedence over sectional and sectarian interests, not, as at present, one in which the national interest claims priority over global needs. And this common interest must be institutionalized in a world organization that can make it effective and is not frustrated by the claims to sovereign rights and sovereignly independent state action. Moral obligation must extend to all living species and to the environment so as to ensure the balance of nature, because on the maintenance of that balance our own survival depends. Similarly, the environment must be included in all economic considerations, and care for it must be taken in all economic practices and economic institutions. To survive, contemporary civilization must change paradigms without further loss of time. The means and the media are already at our disposal in unprecedentedly rapid transport, in the Internet, in satellite communication, and almost instant access to information anywhere in the world and beyond. With such infrastructure all that is needed is the habit of thinking holistically and the will to unite; and the longer the first is delayed and the second is lacking the more precarious our prospects of survival become.

The urgent need for these developments to be met, however, is obstructed by difficulties that appear almost insurmountable. The situation is not simply perilous in terms of terrestrial and meteorological conditions, but also threatens to be psychologically intractable. This aspect of the problem is next to be considered.

NOTES

1. Cf. B. de Spinoza, *Ethics*, Part II, Prop. 13, Schol., Lemma 7, Schol., and Epistle 32 in idem, *Ethics and Selected Letters*, trans. S. Shirley (Indianapolis: HHackett, 1982); E. E. Harris, *Salvation from Despair* (The Hague: Martinus Nijhoff, 1973), chapter 5, 1–3.

2. Cf. W. Leibniz, *Monodology*, trans. R. Latta (Oxford: Clarendon Press, 1898).

3. Cf. Immanuel Kant, *Kritik der reinen Vernünft* (Leipzig: Felix Meiner, 1926), A115–130, B130–140; idem, *Kritik der Urteilskraft* (Leipzig: Felix Meiner, 1924), section 75.

4. Cf. E. E. Harris, *The Spirit of Hegel* (Atlantic Highlands, N.J.: Humanities Press, 1993), chapters 7–8, 10–11.

5. Cf. E. E. Harris, *Hypothesis and Perception* (London: Allen and Unwin, 1970; reprint, Atlantic Highlands, N.J.: Humanities Press, 1996), chapters 7 and 11. Cf. Stuart A. Kauffman, *The Origins of Order* (Oxford: Oxford University Press, 1993); E. E. Harris, *The Foundations of Metaphysics in Science* (London: Allen and Unwin, 1965; reprint, Lanham, Md.: University Press of America, 1983; Atlantic Highlands, N.J.: Humanities Press, 1993), pp. 242–251.

6. Cf. Stuart A. Kauffman, *The Origins of Order* (Oxford: Oxford University Press, 1993); E. E. Harris, *The Foundations of Metaphysics in Science* (London: Allen and Unwin, 1965; reprint, Lanham, Md.: University Press of America, 1983; Atlantic Highlands, N.J.: Humanities Press, 1993), pp. 242–251.

7

Postmodernization—
The Dilemma

The question to be answered is how prevailing habits of think-ing can be freed from latent Newtonian presuppositions and brought into line with the contemporary scientific conceptual outlook. If we call the historical period since the Renaissance "modern," in contrast to the Middle Ages, then we may legiti-mately use the term "postmodern" as applying to the present, post-Einsteinian era. What are the best and most efficient means to the postmodernization of public and political consciousness?

If the twentieth-century scientific paradigm is to permeate the thought of civilized communities throughout the world, and if it takes as long as did that of the Copernican revolution, it would not have spread its influence significantly until the twenty-third century. By that time, if the nations have not yet blown themselves to bits with nuclear bombs or exterminated one another with chemical and biological weapons, the dete-rioration of the environment would have progressed so far that the conditions sustaining human and other life on the planet would have long ceased to exist.

Quite apart from the fact that the influence of a scientific paradigm on other intellectual activities is normally a slow process of infiltration measured in centuries, the present paradigm is especially difficult for the lay public to assimilate because it is almost impossible to imagine in ordinary perceptual terms the concepts entertained by contemporary physicists. The average citizen has difficulty understanding the meaning of a curved space–time, and if a four-dimensional manifold is hard to grasp, still more so is one with yet more dimensions. How does one represent a hyperspherical space curved in four or more dimensions? How does one envisage ten spatiotemporal dimensions rolled up into a ball the size of a proton (as the most advanced super-string theory requires)? The universe, we have learned, is expanding, from a similarly minute volume with an almost infinite mass which exploded with a "big bang" some fifteen thousand million years ago. We have to picture this explosion as unrolling four of the space–time dimensions while leaving the other six self-enfolded. If we can manage to imagine that, we must still be puzzled to locate the point from which it all began. That should surely mark the center of the universe from which the expansion started, but every galaxy is receding from every other with a velocity proportional to the distance, and we are assured that there is no center. Where then was the singularity which initiated the explosion? If we trace back the movement of the galaxies until they all congregate in one conglomeration, will that mark the region at which the universe began its evolution? But at that time the pervasive temperature would have been so high that matter could not have been in a solid state; how then could the universal mass have been concentrated into the reputedly minute space? As space itself exists only as the field of radiant energy, and expands with the energy propelling its contents, when everything is contracted to a point, space itself would have disappeared. Eddington said that where there are no geodesics, space–time shrinks to a point, and for E. A. Milne the spreading light wave creates space: Darkness, he said, is "inimical to geometry."[1] What meaning then can the question have where that point is to be found? Faced with difficulties of conception and imagination like these, the lay person gives up trying to grasp the nature and implications of the scientific theories, so their influence on common habits of thinking is abnormally delayed.

Further, the average person is hardly likely to perceive the philosophical implications of current scientific theories unless guided by philosophers whose training enables them to discern the logical consequences of scientific concepts. It was this that Collingwood identified as the task of the metaphysician. But as long as philosophers reject the possibility of any sort of metaphysics and until they cease to view the world through Newtonian spectacles little progress can be expected.

Appreciation of the essential holism of the contemporary paradigm, however, to which so many eminent physicists have testified, should not have been so difficult, and that ought to have given the lead at least to some philosophers. In the early years of the twentieth century, this seemed to be happening with Whitehead's doctrine affirming the concrescence of prehensions of the entire universe in every actual entity (echoing a similar emphasis on the whole in Leibniz and Hegel); but the trend did not last long and was overwhelmed by the resurrection of Empiricism. The obscurity of some of Whitehead's writing may have deterred his successors from pursuing the route he had chosen; and those comparatively few philosophers who have followed in his footsteps have not stressed as much as they ought the characteristic holism of his doctrine. Neither he nor they sufficiently developed its logical implications.

In any case, the resurgence of logical Empiricism and its subsequent dominance of the field led to a general neglect of the history of philosophy, which should have directed students' attention at least to the holism implicit in Leibniz's philosophy from which so many of Whitehead's ideas seem to have been derived. Further, the convergence of Whitehead's thinking with the Hegelian dialectic, which emerges from what we read in *Adventures of Ideas*, ought to have provided a signpost indicating the future direction of development. That, however, was abruptly terminated by the declaration that all metaphysics was nonsense. The effect of this neglect of the history of philosophy has fully justified Collingwood's assertion that "a philosophy which ignores its own history is a philosophy which spends its labour only to rediscover errors long dead."[2] The errors rediscovered were those committed by the British Empiricists of the seventeenth and eighteenth centuries and their successors, those which led to epistemological and ethical skepticism.

The influence of philosophy on the common outlook and attitude of an age ought to be incisive and pervasive; but we have seen that philosophers in the latter half of the passing century have failed to interpret the contemporary scientific revolution, and their influence upon common thinking, as far as they have exercised any, has been reactionary. Contemporary empiricists in fact early renounced any responsibility for offering guidance to the uninstructed and unreflective. Alfred Ayer declared early on that philosophers had ceased to pontificate on moral or religious matters and had no responsibility for advising politicians. Collingwood's comment in his *Autobiography* is especially telling:

The pupils, whether or not they expected a philosophy that should give them, as that of Green's school had given their fathers, ideals to live for and principles to live by, did not get it; and were told that no philosopher (except of course a bogus philosopher) would even try to give it. The inference which any pupil could draw for himself was that for guidance in the problems of life, since one must not seek it from thinkers or from thinking, from ideals or from principles, one must look to people who were not thinkers (but fools), to processes that were not thinking (but passions), and to rules that were not principles (but rules of expediency). If the realists had wanted to train up a generation of men and women expressly as the potential dupes of every adventurer in morals and politics, commerce or religion, who should appeal to their emotions and promise them private gain which he neither could procure them nor even meant to procure them, no better way of doing it could have been discovered.[3]

It is, perhaps, more to the reactionary thinking of philosophers than to any other cause that the current gap between popular habits of thought and the prevailing scientific paradigm is to be attributed, and what is now required is a revival of reflection upon past achievements in metaphysics and the implications of contemporary scientific ideas.

The root error has been and still is the tendency to think atomistically. At a time when the most advanced scientific theories are indisputably holistic and have recognized that all individuality and self-sufficiency is dependent upon membership in an inclusive whole, the systematic organized structure of which determines the character of its parts, the cast of thought in molds of atomism and separatism that breeds individualism in private and political life and claims to sovereign independence in international affairs is utterly inappropriate. Most if not all of our present ills can be traced back to this separatist tendency. It is, however, scientifically obsolete and has to be renounced.

The first duty of philosophers now must be to interpret the twentieth-century scientific paradigm, to carry out the task which Collingwood prescribed for metaphysics (the disclosure of the absolute presuppositions of science), and to develop their consequences for ethics, economics, and politics. The resulting cast of thought must then percolate through the media, literature, and the Internet to schools and educational institutions, to permeate the approach to practical life and politics of the mass of the people.

But even this, unless it can somehow be accelerated beyond what is normally to be expected, will not be sufficient to extricate humankind from its present predicament, because the process of persuasion needs to be uncommonly rapid if remedies are to be found in time to prevent final disaster. Action has to be taken within years rather than decades and is already much overdue.

The practical dilemma confronting us is that before politicians and social reformers can be persuaded to act, their habits of thought have to be converted from those of the nineteenth century to what they should have been in the twentieth. The process of persuasion is very slow while the need for action is very urgent. The dilemma is further complicated by the fact that even if a more universal ethic were adopted, embracing the biological as well as the human community, and even if the need for sustainable economic development were widely recognized, this enlightenment cannot be translated into practice simply by private persons or nongovernmental organizations without power to legislate; nor can national governments carry out global measures. We have seen that concerted international action is unlikely to be achieved as long as national governments remain sovereign. So the primary requirement is the establishment of a world government. But this will not occur until the Newtonian paradigm has so far lost its influence that the great majority of people throughout the world come to see the necessity for the effective maintenance of a global rule of law and to demand it of their leaders. At the present time this urgent need is recognized by very few. The horse that is needed to pull the cart lags woefully behind its potential driver.

Many writers at the present time, still thinking inveterately in terms of sovereign independence, dismiss the idea of world federation as utopian idealism. The fault of utopianism is that it overlooks the facts and conditions that would make its proposals impracticable. That federalism is not impracticable the various federal republics that have been in existence for centuries give incontrovertable evidence. They have been spectacularly stable and resilient. On the other hand, the utopians are surely those who continue to rely on "law-making treaties" and the undertakings of sovereign states which history has consistently shown to be unreliable, and who cherish the belief that world peace and international justice can be achieved by methods of diplomacy that have signally failed since the sixteenth century, and in the century now about to end have caused unprecedented destruction of life and property in recurrent warfare, have threatened total annihilation in a nuclear catastrophe, and have permitted hideous massacres and untold misery in civil strife.

Unvarying habits of outdated thinking incline the majority of people against world government. Persistence of individualism encourages ideas of freedom from government interference which fail to appreciate the dependence of true liberty on mutual cooperation and the restraint of self-seeking at the expense of others. So many people oppose world federation in the misguided belief that it would involve a world dictatorship with inordinate power. This is a complete misapprehension. In this day and age no world government could result from conquest because that would involve warfare with weap-

erally acceptable, as it is the only political organization that could provide adequate means of solving world problems.

Just as the fathers of the American Constitution had to find ways of defending states' rights while at the same time ensuring central control, so today the nations will need to safeguard their several autonomies while ensuring peace and the rule of law worldwide. What is necessary is to establish a law enforcement system that can punish and restrain individual law breakers without the use of excessive force. Sovereign states cannot be so restrained or punished, especially when they are ruled by irresponsible dictators. Concerning the American colonies' debate about the advantages of federation John Fiske remarked,

When an individual defies the law, you can lock him up in jail, or levy an execution upon his property, and he is helpless as a straw on the billows of the ocean. He cannot raise a militia to protect himself. But when the law is defied by a state, it is quite otherwise. You cannot put a state into jail, nor seize its goods; you can only make war on it, and if you try that expedient you find that the state is not helpless. Its local pride and prejudices are aroused against you, and its militia will turn out in full force to uphold the infraction of the law.

It is high time that the world powers of our own day took heed of this warning.

Why all these considerations have been so widely ignored is probably that the general run of people still think in terms of national isolation, and our problem is how to wean them away from what has become an obsolete conceptual system. Moreover, this has to be done against the clock, for time is running out on the opportunity to save the global ecology.

Perhaps an ideal solution of the problem would be Plato's prescription: to train up philosophers able to recognize the causes and the implications of the present emergency and to elevate them to political leadership, but this is an undertaking which in the present circumstances seems hardly practicable, if for no other reason, because political ascendancy in these days cannot be imposed, as it were, by academic fiat (as Plato seems to have assumed), but, to be acceptable nowadays, it must be by popular election, and it is the popular adoption of the paradigm that presents the primary problem. Certainly the desperate need is for some inspired charismatic leader who has made the philosophical transition and can carry with him or her not only a single nation but the whole international community. That such a leader should arise is devoutly to be wished, but the mere wish cannot produce one.

There are at present some incipient signs of holistic thinking. The Society of Universalists, publishing the journal *Dialogue and Humanism*, and with a fairly farflung membership, raises some hope. The Mundialists centered in Paris, although their membership expands very slowly, excite a certain degree of optimism, and along with them the World Citizens' Assembly organized mainly from Canada and California raises some hopes. The World Constitution and Parliament Association (WCPA), with its Global Ratification and Elections Network (GREN) claiming over thirteen hundred adherent nongovernment organizations with more than 40 million members, is the one organization taking practical steps toward world unity and deserves all the support that can be mustered. Will these organizations thrive with sufficient speed to address the problems? If they could prevail upon the media to publicize their ideas, much might be accomplished by way of molding public opinion; yet again, it is more generally public taste that directs the media than the media that direct public taste, and it is the public attitude that has to be remodeled. At the present time both public and media are steeped in the prevailing habits of Newtonian thinking, and there is little if any sign of a tendency to wean them away from it.

The Universalists' membership comes mostly from among an élite minority, the Mundialists increase in strength at a relative snail's pace, the World Citizens' Assembly has been sidetracked into supporting efforts to establish a popularly elected advisory body that would be an ancillary to the General Assembly of the United Nations. This is a distraction and a waste of energy for the reasons set out earlier: that the proposed assembly (even if it were approved by the Security Council, which is highly doubtful) would have no legislative powers and could exert no real influence upon international negotiations. WCPA and GREN draw support mainly from developing countries whose international influence is minimal and whose internal order is least stable; so there is still a long way to go before any effective progress can be claimed.

At the same time, a widespread world civil society is emerging, indicating the growing recognition of a global common interest. This international civil society is made up of voluntary organizations such as the peace movements, Amnesty International, Greenpeace, Friends of the Earth, the World Wildlife Fund, Oxfam, and the like, as well as the activities of unofficial international groups of politicians such as the Stockholm Initiative. Organizations and movements like these consist of people who sense the universal interest in the causes they advocate and who are mostly motivated by humanitarian and normative considerations, nor are they limited to any one nation.

The development of a world civil society is especially important if world unity is ever to be achieved, but what has so far emerged is not markedly united, nor do the organizations that contribute to it always see very clearly the best way ahead, each pursuing its own particular objective without regard for the others. Failure to understand the cause of the present inefficiency of international organizations has recently been exemplified, for instance, by the Stockholm Initiative, who issued the Report of a Commission on Global Governance which fails dismally to address the needs of the time, because it continues to rely on international institutions that have spectacularly failed for the past fifty years to maintain peace or the rule of law throughout the world. Nor do any other of the major NGOs sufficiently realize the need for worldwide legal support and protection, which cannot be provided by separate independent national governments, yet is required for the rights and conditions that they seek to defend. So, in their endeavor to remain nonpolitical, they eschew any advocacy of world federation.

Civil society can flourish only under the umbrella of a rule of law maintained by an effective law-enforcement system, and this is what is most painfully lacking in the international sphere at the present time. A widespread popular foundation and support for global action is what needs to be developed, such as the NGOs previously mentioned might generate, an approach that would consolidate into the desired global outlook required for world union. Their efforts, however, need to be coordinated and concerted behind a single comprehensive aim, which should include propaganda and campaigning for effective democratic world government, and because their leaders still think in Newtonian terms, they are generally averse to any suggestion that they should advocate world unity.

Yet again, in order to galvanize the organizations that promise to generate the appropriate world civil society into the necessary coordination and concerted action, they must be somehow induced to think holistically, not just about their own particular aims, but equally about the interdependence of these aims with other global objectives; and the still unanswered question is how this can be achieved.

Notably today, many writers and thinkers who are aware of the problems besetting the human race are resolutely opposed to the idea of world government. They reject it out of hand without reason or argument. This was the case with the Commission on Global Governance of the Stockholm Initiative, who quite illogically denounced world government in their report as undemocratic, "accommodating to power, more hospitable to hegemonic ambition, and more reinforcing of the roles of states and governments rather than the rights of people."[5]

The same antagonism to world federalism is apparent in the work of Al Gore in his book *Earth in the Balance*, despite what he admitted in the passage quoted previously. A similar attitude is taken by Professor Richard Falk, who dismisses the idea of world federation as merely academic and, as he puts it, "rational."[6] Why rationality should be considered a disadvantage is not obvious. It is perhaps more easily understandable that politicians would be so rigidly averse to the contemplation of any such idea, because they imagine that they have something important to lose, namely power. But in this they are wrong, for their legitimate power would be as accessible in a world federation as it is at present. As these persons are all highly intelligent, one can attribute their opposition to the idea of federal unity and their continued devotion to sovereign independence, in the last resort, only to their continual habit of thinking in the mold of the Newtonian paradigm.

Consequently, we are confronted not only by stupendous problems of peacekeeping, world administration, sustainable production, and environmental conservation, but we are also impaled on the horns of the dilemma: how on the one hand to direct academic thinking and to school public opinion, which inevitably is a slow process, and on the other to prompt the necessary action which depends on the prior conversion of public opinion in time to avert imminent catastrophe.

The main responsibility falls to philosophers and philosopher–scientists who are able to grasp and appreciate the sociopolitical importance of the new paradigm to do everything possible to revolutionize popular thinking and to persuade politicians of the danger of continuing to think in what have now become outdated categories. They have at their disposal unprecedented methods of mass communication, which can be used along with traditional educational media to instruct and persuade both the young and the mature that atomism of every kind and application has to be sloughed off and that the habit of thinking in terms of the whole has to be acquired and cultivated.

Once such habits of thought have been widely inculcated, average intelligence would surely be enough to enable people to draw the obvious conclusions from available evidence and become sufficiently aware of the imminent dangers overshadowing the human race to pressure their political leaders into taking the appropriate action. It is also of the utmost importance that these leaders themselves should recognize the futility of acting merely within national boundaries and the imperative necessity for globally concerted action, giving priority to the interests of humanity as a whole.

Amid all the enthusiasm generated in preparation for the coming millennial year, little is apparent that will address glo-

bal problems adequately or swiftly. The major millennium cel-
ebration should be marked, not so much by colossal monu-
ments and giant domes as by the initiation of a world movement
toward social, economic, and political unity inspired by a new
philosophical analysis consonant with contemporary science.
This ought to be the primary millennial project—but as yet there
is little or no sign of anything of the sort.

Possibly, to encourage this kind of movement, some expert
and professional convention such as Pugwash could be as-
sembled to report on world conditions and the action needed
to address problems and promote survival. Once again, the right
kind of leadership is required to bring this about. Emphasis
should be laid on the action needed, whereas in the past, while
scientific conferences have reported on the progressing destruction of the environment, little has been said about what measures should be taken to reverse the deleterious processes.
Moreover, when this has been set out, the genuinely efficient
political and legal means of enactment and enforcement have
yet to be instituted.

If human and other living beings are to survive the coming
century, it is essential that we should learn to think holistically.
The twentieth-century scientific paradigm must become intrinsic to the millennial outlook, and the millennial objective ought
to be the initiation of unified global organization. The unity of
humanity should be the watchword of the new epoch, inspir-
ing all our thinking and action. It is essential to stress the unity
of the whole in and through difference. In all local action the
global perspective has to be kept firmly in mind.

NOTES

1. Cf. Arthur Eddington, *Space, Time and Gravitation* (Cambridge:
Cambridge University Press, 1935), pp. 157f; and E. A. Milne, *Proceedings of the Royal Society of Edinburgh*, section A, vol. 62 (1943–1944).

2. R. G. Collingwood, "Croce's Philosophy of History," *The Hibbert
Journal* (1921), reprinted in *Essays in the Philosophy of History*, ed. W.
Debbins (Austin: University of Texas Press, 1965), p. 4.

3. R. G. Collingwood, *An Autobiography* (Oxford: Oxford University Press, 1940), p. 48.

4. Al Gore, *Earth in the Balance* (Boston: Houghton Mifflin, 1992), p. 204.

5. Cf. Commission on Global Governance, *Our Global Neighbour-hood* (Oxford: Oxford University Press, 1995), p. xvi.

6. See Richard A. Falk, *Explorations at the Edge of Time* (Philadelphia: Temple University Press, 1992).

Select Bibliography

Alexander, S. *Space, Time and Deity*. London: Macmillan, 1920.

Aristotle. *Metaphysics*. Trans. H. Tredennick. Cambridge: Harvard University Press, 1957.

———. *Physics*. Trans. F. M. Cornford and P. H. Wicksteed. Cambridge: Harvard University Press, 1935.

Austin, J. *Lectures on Jurisprudence*. London: Murray, 1913.

Ayer, A. J. *The Foundations of Empirical Knowledge*. London: Macmillan, 1951.

———. *Language, Truth and Logic*. London: Macmillan, 1946.

———. *The Problem of Knowledge*. Harmondsworth, England: Penguin, 1956.

Baier, K. *The Moral Point of View*. Ithaca, N.Y.: Cornell University Press, 1958.

Barker, E. *Principles of Social and Political Theory*. Oxford: Clarendon Press, 1951.

Barrow, J. D., and F. J. Tipler. *The Anthropic Cosmological Principle*. Oxford: Oxford University Press, 1988.

Bates, M. *The Forest and the Sea*. New York: Random House, 1960.

Bergson, H. *Creative Evolution*. London: Macmillan, 1911.

———. *L'Evolution Creatrice*. Paris: Librairie Felix Alcan, 1918.

Berkeley, G. *A New Theory of Vision: Principles of Human Knowledge; Three Dialogues between Hylas and Philonous.* London: Scribner, 1929.

Blackstone, W. *Commentaries on the Laws of England.* London: Butterworth, 1825.

Bohm, D. *Wholeness and the Implicate Order.* London: Routledge and Kegan Paul, 1980.

Born, M. *Einstein's Theory of Relativity.* New York: Dover, 1965.

Bosanquet, B. *The Philosophical Theory of the State.* London: Macmillan, 1925.

————. *The Principle of Individuality and Value.* London: Macmillan, 1927.

Bricker, P., and R.I.G. Hughes. *Philosophical Perspectives on Newtonian Science.* Cambridge, Mass.: MIT Press, 1990.

Capra, F. *The Tao of Physics.* London: Fontana, 1985.

Collingwood, R. G. *An Autobiography.* Oxford: Oxford University Press, 1940.

————. *An Essay on Metaphysics.* Oxford: Clarendon Press, 1940.

————. *An Essay on Philosophical Method.* Oxford: Clarendon Press, 1933, 1950.

————. *Essays in the Philosophy of History.* Ed. W. Debbins. Austin: University of Texas Press, 1965.

————. *The Idea of Nature.* Oxford: Clarendon Press, 1945.

————. *Speculum Mentis.* Oxford: Clarendon Press, 1924.

Commission on Global Governance. *Our Global Neighbourhood.* Oxford: Oxford University Press, 1995.

Cornford, F. M. *From Religion to Philosophy: A Study of the Origins of Western Speculation.* New York: Harper Torchbooks, 1957.

Curry, W. B. *The Case for Federal Union.* Harmondsworth, England: Penguin, 1939.

Curtis, L. *World War, Its Cause and Cure.* London: Oxford University Press, 1945.

Darwin, C. *The Origin of Species.* London: Watts, 1929.

Davies, P. *God and the New Physics.* London: Dent, 1984, 1986.

de Broglie, L. *The Revolution in Physics.* Trans. R. W. Wiemeyer. London: Routledge and Kegan Paul, 1954.

de Santillana, G. *The Origins of Scientific Thought.* New York: Mentor, 1961.

Descartes, R. *The Philosophical Works of Descartes*, vol. 1. Ed. E. S. Haldane and G.R.T. Ross. Cambridge: Cambridge University Press, 1931.

Dennett, D. *Consciousness Explained*. Boston: Back Bay, 1991.
————. *Darwin's Dangerous Idea: Evolution and the Meanings of Life*. New York: Simon and Schuster, 1995.
Driesch, H. *The Problem of Individuality*. London: Macmillan, 1927.
————. *Science and Philosophy of the Organism*. London: A. & C. Black, 1927.
Eddington, A. *The Expanding Universe*. Cambridge: Cambridge University Press, 1933.
————. *The Nature of the Physical World*. Cambridge: Cambridge University Press, 1928–1948.
————. *The Philosophy of Physical Science*. Cambridge: Cambridge University Press, 1939.
————. *Space, Time and Gravitation*. Cambridge: Cambridge University Press, 1935.
Einstein, A. *Relativity: The Special and General Theory*. New York: Crown, 1961.
Einstein, A., and L. Infeld. *The Evolution of Modern Physics*. New York: Simon and Schuster, 1954.
Falk, R. *Explorations at the Edge of Time*. Philadelphia: Temple University Press, 1992.
Flew, A. *God and Philosophy*. London: Hutchinson, 1961.
Flew, A., and A. MacIntyre. *New Essays in Philosophical Theology*. New York: Macmillan, 1955.
Foucault, M. *The Archeology of Knowledge*. London: Tavistock, 1972–1986.
Freud, S. *Civilization and Its Discontents*. Trans. J. Trachey. London: Hogarth Press, 1963.
————. *Totem and Taboo*. Trans. J. Strachey. London: Hogarth Press, 1950.
Gleick, J. *Chaos*. New York: Viking, 1987.
Gore, A. *Earth in the Balance*. Boston: Houghton Mifflin, 1992.
Hamilton, A., J. Jay, and J. Madison. *The Federalist*. New York: Random House, 1940.
Hammond, A., ed. *Environmental Almanac*. World Resources Institute. Boston: Houghton Mifflin, 1992.
Harris, E. E. *Atheism and Theism*. Atlantic Highlands, N.J.: Humanities Press, 1993.
————. *Cosmos and Anthropos*. Atlantic Highlands, N.J.: Humanities Press, 1991.
————. *Cosmos and Theos*. Atlantic Highlands, N.J. : Humanities Press, 1992.

————. "Darwinism and God." *The International Philosophical Quarterly* 34, no. 3 (September 1999).

————. *Formal, Transcendental and Dialectical Thinking.* Albany: State University of New York Press, 1987.

————. *The Foundations of Metaphysics in Science.* London: Allen and Unwin, 1965. Reprint, Lanham, Md.: University Press of America, 1983; Atlantic Highlands, N.J.: Humanities Press, 1993.

————. *Hypothesis and Perception.* London: Allen and Unwin, 1970. Reprint, Atlantic Highlands, N.J.: Humanities Press, 1996.

————. "Natural Law and Naturalism" (Suarez lecture). *The International Philosophical Quarterly* 23, no. 2 (June 1983).

————. *One World or None.* Atlantic Highlands, N.J.: Humanities Press, 1993.

————. *The Spirit of Hegel.* Atlantic Highlands, N.J.: Humanities Press, 1993.

————. *The Substance of Spinoza.* Atlantic Highlands, N.J.: Humanities Press, 1995.

————. *The Survival of Political Man.* Johannesburg: Witwatersrand University Press, 1950.

Harris, E. E., and J. Yunker, eds. *Toward Genuine Global Governance: Critical Reactions to "Our Global Neighborhood."* Westport, Conn.: Praeger, 1999.

Harris, J. M. *World Agriculture and the Environment.* New York: Garland, 1990.

Harris, J. M., and Anne-Marie Codur. *Macro-Economics and the Environment.* Medford, Mass.: Tufts University Press, 1998.

Hawking, S. *A Brief History of Time: From the Big Bang to Black Holes.* New York: Bantam, 1988.

Hegel, G.W.F. *Hegel's Phenomenology of Spirit.* Trans. A. V. Miller. Oxford: Clarendon Press, 1977.

————. *Hegel's Philosophy of Right.* Trans. T. M. Knox. Oxford: Clarendon Press, 1953.

————. *Hegel's Science of Logic.* Trans. A. V. Miller. New York: Humanities Press, 1969.

————. *Werke.* 20 vols. Frankfurt-am-Main: Suhrkamp Verlag, 1971.

Heisenberg, W. *Philosophic Problems of Nuclear Science.* London: Faber and Faber, 1934.

————. *Physics and Philosophy.* New York: Harper, 1958, 1962.

Hobbes, T. *Leviathan.* Oxford: Clarendon Press, 1943.

Hume, D. *Enquiry Concerning the Human Understanding.* Ed. L. A. Selby-Bigge. Oxford: Clarendon Press, 1902, 1955.

————. *A Treatise of Human Nature.* Ed. L. A. Selby-Bigge. Oxford: Clarendon Press, 1888.

Husserl, E. *The Crisis of European Sciences.* Trans. D. Carr. Evanston, Ill.: Northwestern University Press, 1970.

Kaku, M., and J. Trainer. *Beyond Einstein: The Cosmic Quest for the Theory of the Universe.* New York: Bantam, 1987.

Kant, I. *Kritik der praktischen Vernünft.* Leipzig: Felix Meiner, 1920.

————. *Kritik der reinen Vernünft.* Leipzig: Felix Meiner, 1926.

————. *Kritik der Urteilskraft.* Leipzig: Felix Meiner, 1924.

Kauffman, S. A. *The Origins of Order.* Oxford: Oxford University Press, 1993.

Keeton, G. *National Sovereignty and International Order.* London: Stevens, Peace Book, 1939.

Kitchener, R., ed. *The World View of Contemporary Physics: Does It Need a New Metaphysics?* Albany: State University of New York Press, 1988.

Kuhn, T. S. *The Structure of Scientific Revolutions.* Chicago: University of Chicago Press, 1962–1964.

Latta, R. *Leibniz: The Monadology.* Oxford: Clarendon Press, 1898.

Lauterpacht, H. *The Function of Law in the International Community.* Oxford: Clarendon Press, 1933.

Locke, J. *An Essay Concerning Human Understanding.* Ed. A. C. Fraser. Oxford: Clarendon Press, 1894.

————. *Of Civil Government: Two Treatises.* London: Dent, 1924.

MacIntyre, A. *After Virtue.* Notre Dame, Ind.: Notre Dame University Press, 1981.

Marx, K. *Capital.* Trans. G.D.H. Cole. London: Dent, 1933.

————. *Selected Works.* 2 vols. Ed. V. Adoratsky. London: Lawrence and Wishart, 1942–1945.

Mill, J. S. *Theism.* New York: Bobbs-Merrill, 1957.

————. *Three Essays on Religion, Theism.* London: Longman, 1875.

————. *Utilitarianism, Liberty, Representative Government.* London: Dent, 1910–1922.

Montesquieu, C. L. *L'esprit de lois.* Paris, 1927.

Moore, G. E. *Principia Ethica,* Cambridge: Cambridge University Press, 1903.

Newton, I. *Principia Mathematica.* Trans. A. Motte. Berkeley and Los Angeles: University of California Press, 1966.

Nietzsche, F. *Beyond Good and Evil.* Trans. R. J. Hollingdale. Harmondsworth, England: Penguin, 1973–1974.

————. *Thus Spake Zarathustra.* Trans. R. J. Hollingdale. Harmondsworth, England: Penguin, 1961–1974.

————. *Twilight of the Idols: The Anti-Christ*. Trans. R. J. Hollingdale. Harmondsworth, England: Penguin, 1972.

Nozick, R. *Anarchy, State, and Utopia*. New York: Basic Books, 1974.

Planck, M. *The Universe in the Light of Modern Physics*. London: Allen and Unwin, 1937.

————. *Where Is Science Going?* London: Allen and Unwin, 1933.

Plato. *Republic*. Trans. A. Bloom. New York: Basic Books, 1968.

————. *Timaeus*. Trans. F. M. Cornford. In *Plato's Cosmology*. London: Routledge and Kegan Paul, 1948.

Prigogine, I., and G. Nicolis. *Self-Organization in Non-Equilibrium Systems*. New York: Wiley, 1977.

Rawls, J. *A Theory of Justice*. Cambridge: Harvard University Press, 1971.

Reichenbach, H. *Experience and Prediction*. Chicago: University of Chicago Press, 1938.

————. *The Rise of Scientific Philosophy*. Berkeley and Los Angeles: University of California Press, 1951.

Ross, W. D. *Foundations of Ethics*. Oxford: Clarendon Press, 1939.

Rousseau, J. J. *Du contrat social*. Leipzig: Gerhard Fleischer, 1818.

————. *Social Contract*. London: Oxford University Press, 1945.

Russell, B. *Has Man a Future?* Harmondsworth, England: Penguin, 1961.

————. *Human Knowledge*. London: Allen and Unwin, 1948.

————. *The Problems of Philosophy*. London: Oxford University Press, 1950.

Sartre, J. *Being and Nothingness*. Trans. H. E. Barnes. New York: Washington Square Press, 1966, 1969.

Schilpp, P. A., ed. *Einstein: Philosopher–Scientist*. New York: Harper Torchbooks, 1959.

————. *The Philosophy of Bertrand Russell*. Evanston, Ill.: Open Court, 1946.

Schwartzenberger, G. *Power Politics*. London: Stevens & Sons, 1951.

Sciama, D. W. *The Unity of the Universe*. New York: Doubleday Anchor, 1961.

Smith A. *The Wealth of Nations*. New York: Random House, 1937.

Snell, B. *The Discovery of the Mind*. New York: Harper Torchbooks, 1960.

Spencer, H. *Man vs. the State*. London: William and Norgate, 1897.

Spinoza, B. de *Collected Works*. Trans. and ed. E. Curley. Princeton, N.J.: Princeton University Press, 1985.

————. *The Ethics and Selected Letters*. Trans. S. Shirley. Indianapolis: Hackett, 1982.

————. *Opera*, in *Auftrag der Heidelberger Akademie der Wissenschaften* (Publications of the Heidelberg Academy of Sciences). Ed. Karl Gebhardt. Heidelberg: Winter, 1925, 1972.

Stonier, T. *Nuclear Disaster*. Harmondsworth, England: Penguin, 1963–1964.

Streit, C. *Union Now*. New York: Harper, 1939.

Teilhard de Chardin, P. *The Phenomenon of Man*. Trans. B. Wall. New York: Harper and Row, 1959.

Thomas, L. *The Lives of a Cell*. New York: Viking, 1974.

Weare, K. C. *Federal Government*. London, New York: Oxford University Press, 1946.

Whitehead, A. N. *Adventures of Ideas*. Cambridge: Cambridge University Press, 1942.

————. *Process and Reality*. Cambridge: Cambridge University Press, 1929.

————. *Science and the Modern World*. Cambridge: Cambridge University Press, 1930.

Wittgenstein, L. *Philosophical Investigations*. Trans. G.E.M. Anscombe. Oxford: Blackwell, 1968.

————. *Tractatus logico-philosophicus*. London: Routledge and Kegan Paul, 1922, 1947.

Worldwatch. *State of the World—1992*. London: W. W. Norton, 1992–1997.

Index

India, 15
Individualism, 31–33, 51, 56, 72, 90, 103, 106, 121
Individuality, 105
Indonesia, 52
Induction, 29, 41
Inertia, 11, 13, 100
Infeld, Leopold, 75, 77, 86
Information, 115
Initial conditions, 79
Intention, 98
Interest: American, 61; British national, 61; common, 56, 61, 115, 124; national, 61, 63, 64–66, 70–71, 108, 115; vested, 106; vital, 61
Internet, 115, 121
Intuition, 23
Ion, 86
Iraq, 46, 64, 65, 67
Israel, 45, 64, 67

Jacobi, F. H., 36
Jahwe, 12
John, St., 112
Judiciary, 32
Jupiter, 79
Justice, 51; International Court of, 61, 66; Permanent Court of International, 61, 62; social, 51

Kant, Immanuel, 29–30, 36, 39–40, 43, 47, 99–100
Kantianism, 42
Kauffman, Stuart, 83
Kavorno-Karabakh, 65
Keeton, George, 71
Kepler, Johannes, 11, 19
Kierkegaard, Søren, 47
Knowledge, problem of, 22
Korea, 65

Kosovo, 61, 65, 109
Kuhn, Thomas, 1–2, 9
Kuwait, 65

Labor, 34, 107; division of, 105
Laplace, Pierre Simon, 36
Latin America, 60
Lauterpacht, H., 62
Law, 5–6, 56, 58, 105; civil, 6; divine, 105; International, 57, 60–64; moral, 30; of nature, 6, 13, 30–31, 57, 103; of reason, 6, 13; rule of, 65, 109, 122–123, 127, 129
League of Nations, 60, 62–63, 67
Legislature, 32
Leibniz, Wilhelm von Gottfried, 24, 99, 119
Leptons, 87
Leucippus, 5, 8, 14
Liberty, 37, 122
Library of Mendel, 47
Life, 83–84; force, 94; intelligent, 83–84, 99, 103, 112
Light, velocity of, 76
Lloyd-Morgan, C., 94
Locke, John, 25–26, 31–32, 56, 72; *An Essay Concerning Human Understanding*, 24
Logic, 93–99; dialectical, 10, 97; extensional, 97; formal, 41; of system, 95–96; symbolic, 41
Logos, 112
Lorenz, Edward, 79
Louis XIV of France, 63
Love, 5
Lovelock, James E., 82
Lyceum, 5

MacDonald, Margaret, 44
Macromolecules, 87–89

ABOUT THE AUTHOR

Errol E. Harris is Professor of Philosophy Emeritus at Northwestern University. The author of 25 books on philosophy, Professor Harris has held named chairs at four universities and has taught worldwide. His latest book, coedited with James A. Yunker, is *Toward Genuine Global Governance: Critical Reactions to "Our Global Neighborhood"* (Praeger, 1999).

ISBN 0-275-96830-8

EAN

9 780275 968304

HARDCOVER BAR CODE